Lecture Notes in Computer Science 10857

Commenced Publication in 1973
Founding and Former Series Editors:
Gerhard Goos, Juris Hartmanis, and Jan van Leeuwen

More information about this series at http://www.springer.com/series/7412

Maria De Marsico · Gabriella Sanniti di Baja
Ana Fred (Eds.)

Pattern Recognition Applications and Methods

6th International Conference, ICPRAM 2017
Porto, Portugal, February 24–26, 2017
Revised Selected Papers

 Springer

Editors
Maria De Marsico
Sapienza Università di Roma
Rome
Italy

Ana Fred
University of Lisbon
Lisbon
Portugal

Gabriella Sanniti di Baja
ICAR-CNR
Naples, Napoli
Italy

ISSN 0302-9743   ISSN 1611-3349   (electronic)
Lecture Notes in Computer Science
ISBN 978-3-319-93646-8   ISBN 978-3-319-93647-5   (eBook)
https://doi.org/10.1007/978-3-319-93647-5

Library of Congress Control Number: 2018947324

LNCS Sublibrary: SL6 – Image Processing, Computer Vision, Pattern Recognition, and Graphics

Printed on acid-free paper

This Springer imprint is published by the registered company Springer International Publishing AG
part of Springer Nature
The registered company address is: Gewerbestrasse 11, 6330 Cham, Switzerland

Preface

The 13 chapters of this book are the extended and revised versions of selected papers presented at the 6th International Conference on Pattern Recognition Applications and Methods (ICPRAM 2017), held in Porto, Portugal, February 24–26, 2016. Since its first edition, the purpose of the ICPRAM conference has been to establish and strengthen contacts among researchers active in different research fields related to pattern recognition in its wider connotation, both from theoretical and application perspectives. This book collects the best contributions along this line. In particular, they represent the most interesting and relevant part of all submissions (11%) received for ICPRAM 2017. The pre-selection process was carried out by the general chair and the program chairs of the event by taking into account a number of criteria such as classifications and comments provided by the ICPRAM 2017 Program Committee members, the session chairs' assessment of presentation quality, and the program chairs' global view of all papers included in the technical program. Then, the authors of the pre-selected papers were invited to submit a revised and extended version of their work; a new reviewing process was performed to check whether the submitted extended papers were characterized by a sufficient amount of innovative material, with respect to the discussion of the proposed approaches, the presentation of theoretical as well as operational details, and experiments.

We believe that this book can contribute to the understanding of relevant trends of current research on pattern recognition in the areas covered by the collected papers. As with the conference organization, we divided the papers into two main tracks: "Applications" and "Methods." The five papers dealing with methods are presented first, not because they are more important but because they have a more general scope, and each of them may offer inspiration for different applications. Then the eight papers presenting a wide variety of applications follow.

The papers dealing with methods are introduced next.

In "Control Variates as a Variance Reduction Technique for Random Projections," by Keegan Kang and Giles Hooker, control variates are used as a variance reduction technique in Monte Carlo integration, making use of positively correlated variables to bring about a reduction of variance for estimated data.

In "Graph Classification with Mapping Distance Graph Kernels," by Tetsuya Kataoka, Eimi Shiotsuki, and Akihiro Inokuchi, two novel graph kernels are proposed, namely, a mapping distance kernel with stars (MDKS), and mapping distance kernel with vectors (MDKV), to classify labeled graphs more accurately than existing methods.

In "Domain Adaptation Transfer Learning by Kernel Representation Adaptation," by Xiaoyi Chen and Régis Lengellé, a new SVM-based approach with a supplementary maximum mean discrepancy (MMD)-like constraint is proposed, as well as a kernel principal component analysis (KPCA)-based transfer learning method. Both methods are compared with other transfer learning methods from the literature to show their efficiency on synthetic and real datasets.

In "Optimal Linear Imputation with a Convergence Guarantee," by Yehezkel S. Resheff and Daphna Weinshall, a method for imputation of missing values is proposed, which is guaranteed to converge to a local minimum. The performance of the method is shown to be markedly superior in comparison with other methods.

The paper "Condensing Deep Fisher Vectors: To Choose or to Compress?" by Sarah Ahmed and Tayyaba Azim shows that feature compression is a better choice than feature selection for reducing data high-dimensional memory. In particular, this holds when dealing with large-scale retrieval of high dimensional Fisher vectors, when they are derived from deep or shallow stochastic models such as restricted Boltzmann machine.

The group of papers dealing with Applications follows.

The paper "Emotion Recognition Using Neighborhood Components Analysis and ECG/HRV-Based Features," by Hany Ferdinando, Tapio Seppänen, and Esko Alasaarela, explores how much neighborhood component analysis (NCA) enhances emotion recognition using ECG-derived features. Results with the MAHNOB-HCI database were validated using subject-dependent and subject-independent scenarios with kNN as classifier for 3-class problem in valence and arousal.

In "A Conversive Hidden Non-Markovian Model Approach for 2D and 3D Online Movement Trajectory Verification," by Tim Dittmar, Claudia Krull, and Graham Horton, an approach for stochastic modelling of movement trajectories is presented, where the models are based on conversive hidden non-Markovian models. A verification system is presented that creates trajectory models from several examples. Its performance is deduced from experiments on different data sets including signatures, doodles, pseudo-signatures, and hand gestures recorded with a Kinect.

The paper "Prediction of User Interest by Predicting Product Text Reviews," by Esteban García-Cuesta, Daniel Gómez-Vergel, Luis Gracia-Expósito, José Manuel López-López, and María Vela-Pérez, deals with shopping websites providing social network services to collect the opinions of the users on items available for purchasing. A prediction is done based on the sets of words that users would use should they express their opinions and interests on items not yet reviewed. To this aim, careful attention is given to the internal consistency of the model by relying on well-known facts of linguistic analysis, collaborative filtering techniques, and matrix factorization methods.

In the paper "Blood Vessel Delineation in Endoscopic Images with Deep Learning Based Scene Classification," by Mayank Golhar, Yuji Iwahori M. K. Bhuyan, Kenji Funahashi, and Kunio Kasugai, a novel blood vessel extraction methodology is proposed. First, a high-level classification of the input endoscopic images into four classes is carried out. Then, the classified images containing blood vessel information are processed with a Frangi vesselness filter. The results of the proposed blood vessel delineation algorithm were found to give better accuracy than the vanilla Frangi vesselness filter and the BCOSFIRE filter, increasing it by 8% and 5%, respectively.

In "Semi-Automated Testing of an Architectural Floor Plan Retrieval Framework: Quantitative and Qualitative Comparison of Semantic Pattern-Based Matching Approaches," by Qamer Uddin Sabri, Johannes Bayer, Viktor Ayzenshtadt,

Syed Saqib Bukhari, Klaus-Dieter Althoff, and Andreas Dengel, case-based reasoning and (in)exact graph matching are utilized to construct an end-to-end system for floor plan retrieval, accessible by a refined version of a design-supporting Web interface. A floor plan is modeled as a graph, where each room is represented as a node and the relations between rooms are modeled as edges.

In "Characterization of a Virtual Glove for Hand Rehabilitation Based on Orthogonal LEAP Controllers," by Giuseppe Placidi, Luigi Cinque, Matteo Polsinelli, and Matteo Spezialetti, a multi-sensor approach, namely, the virtual glove (VG), is presented. It is based on the simultaneous use of two orthogonal LEAP motion controllers. An engineered version of the VG is described, and its characterization is performed through spatial measurements.

The paper "Congestion Analysis Across Locations Based on Wi-Fi Signal Sensing," by Atsushi Shimada, Kaito Oka, Masaki Igarashi, and Rin-ichiro Taniguchi, deals with congestion analysis focusing on perceptual congestion rather than on objective, quantitative congestion. The relationship between quantitative and perceptual congestion is also analyzed. To this aim, a system for estimating and visualizing congestion and collecting user reports about congestion is described.

In "Text Line Segmentation in Handwritten Documents Based on Connected Components Trajectory Generation," by Insaf Setitra, Abdelkrim Meziane, Zineb Hadjadj, and Nawfel Bengherbia, a novel approach of text line segmentation based on tracking is presented. Each connected component is considered as a moving object along its respective line, and finds its best match given its history motion, i.e., the closest connected component that lies in its trajectory.

Finally, we would like to express our gratitude to all the authors for their contributions, and to the reviewers, who helped ensure the quality of this book. Our thanks are also due to the INSTICC staff who supported both the conference and the preparation of this book.

February 2017

Maria De Marsico
Gabriella Sanniti di Baja
Ana Fred

Organization

Conference Chair

Ana Fred   Instituto de Telecomunicações, IST, Portugal

Program Co-chairs

Maria De Marsico   Sapienza Università di Roma, Italy
Gabriella Sanniti di Baja   Italian National Research Council CNR, Italy

Program Committee

Andrea F. Abate   University of Salerno, Italy
Ashraf AbdelRaouf   Misr International University MIU, Egypt
Rahib Abiyev   Near East University, Turkey
Mayer Aladjem   Ben-Gurion University of the Negev, Israel
Rocío Alaiz-Rodríguez   Universidad de Leon, Spain
Andrea Albarelli   Università Ca' Foscari Venezia, Italy
Guillem Alenya   Institut de Robòtica i Informàtica Industrial,
                                    CSIC-UPC, Spain
Luís Alexandre   UBI/IT, Portugal
Kevin Bailly   Pierre and Marie Curie University, France
Gabriella Sanniti di Baja   Italian National Research Council CNR, Italy
Emili Balaguer-Ballester   Bournemouth University, UK
Enrique Munoz Ballester   Università degli Studi di Milano, Italy
Jorge Batista   Institute of Systems and Robotics, Portugal
Stefano Berretti   University of Florence, Italy
Monica Bianchini   University of Siena, Italy
Battista Biggio   University of Cagliari, Italy
Nizar Bouguila   Concordia University, Canada
Francesca Bovolo   Fondazione Bruno Kessler, Italy
Paula Brito   Universidade do Porto, Portugal
Samuel Rota Bulò   Fondazione Bruno Kessler, Italy
Javier Calpe   Universitat de València, Spain
Francesco Camastra   University of Naples Parthenope, Italy
Ramón A. Mollineda   Universitat Jaume I, Spain
  Cárdenas
Marco La Cascia   Università degli Studi di Palermo, Italy
Michelangelo Ceci   University of Bari, Italy
Mehmet Celenk   Ohio University, USA
Jocelyn Chanussot   Grenoble Institute of Technology, France
Amitava Chatterjee   Jadavpur University, India

Chi Hau Chen	University of Massachusetts Dartmouth, USA
SongCan Chen	Nanjing University of Aeronautics and Astronautics, China
Dmitry Chetverikov	MTA SZTAKI, Hungary
Jen-Tzung Chien	National Chiao Tung University, Taiwan
Ioannis Christou	Athens Information Technology, Greece
Francesco Ciompi	Radboud University Medical Center, The Netherlands
Miguel Coimbra	University of Porto, Portugal
Sergio Cruces	Universidad de Sevilla, Spain
Duc-Tien Dang-Nguyen	Dublin City University, Ireland
Yago Diez	Yamagata University, Japan
Jean-Louis Dillenseger	Université de Rennes 1, France
Junyu Dong	Ocean University of China, China
Mahmoud El-Sakka	The University of Western Ontario, Canada
Kjersti Engan	University of Stavanger, Norway
Haluk Eren	Firat University, Turkey
Yaokai Feng	Kyushu University, Japan
Mário Figueiredo	Instituto de Telecomunicações, IST, University of Lisbon, Portugal
Gernot A. Fink	TU Dortmund, Germany
Ana Fred	Instituto de Telecomunicações, IST, Portugal
Diamantino Freitas	Universidade do Porto, Portugal
Muhammad Marwan Muhammad Fuad	Technical University of Denmark, Denmark
Giorgio Fumera	University of Cagliari, Italy
Vicente Garcia	Autonomous University of Ciudad Juarez, Mexico
James Geller	New Jersey Institute of Technology, USA
Angelo Genovese	Università degli Studi di Milano, Italy
Giorgio Giacinto	University of Cagliari, Italy
Eric Granger	École de Technologie Supérieure, Canada
Marcin Grzegorzek	University of Siegen, Germany
Sébastien Guérif	University Paris 13, SPC, France
Michal Haindl	Institute of Information Theory and Automation, Czech Republic
Barbara Hammer	Bielefeld University, Germany
Makoto Hasegawa	Tokyo Denki University, Japan
Pablo Hennings-Yeomans	Sysomos, Canada
Laurent Heutte	Université de Rouen, France
Kouichi Hirata	Kyushu Institute of Technology, Japan
Sean Holden	University of Cambridge, UK
Su-Yun Huang	Academia Sinica, Taiwan
Jose M. Iñesta	Universidad de Alicante, Spain
Yuji Iwahori	Chubu University, Japan
Sarangapani Jagannathan	Missouri University of Science and Technology, USA
Lisimachos Kondi	University of Ionnina, Greece
Mario Köppen	Kyushu Institute of Technology, Japan

Constantine Kotropoulos	Aristotle University of Thessaloniki, Greece
Sotiris Kotsiantis	University of Patras, Greece
Konstantinos Koutroumbas	National Observatory of Athens, Greece
Kidiyo Kpalma	INSA de Rennes, France
Marek Kretowski	Bialystok University of Technology, Poland
Adam Krzyzak	Concordia University, Canada
Piotr Kulczycki	Polish Academy of Sciences, Poland
Shi-wook Lee	National Institute of Advanced Industrial Science and Technology, Japan
Young-Koo Lee	Kyung Hee University, South Korea
Jochen Leidner	Thomson Reuters, UK
Boaz Lerner	Ben-Gurion University of the Negev, Israel
Aristidis Likas	University of Ioannina, Greece
Andrzej Lingas	Lund University, Sweden
Shizhu Liu	Apple, USA
Eduardo Lleida	Universidad de Zaragoza, Spain
Gaelle Loosli	Clermont Université, France
Alessandra Lumini	Università di Bologna, Italy
Juan Luo	George Mason University, USA
Francesco Marcelloni	University of Pisa, Italy
Elena Marchiori	Radboud University, The Netherlands
Gian Luca Marcialis	Università degli Studi di Cagliari, Italy
Urszula Markowska-Kaczmar	Wroclaw University of Technology, Poland
Maria De Marsico	Sapienza Università di Roma, Italy
Sally Mcclean	University of Ulster, UK
Hongying Meng	Brunel University London, UK
Domingo Mery	Universidad Catolica de Chile, Chile
Alessio Micheli	University of Pisa, Italy
Delia Alexandrina Mitrea	Technical University of Cluj-Napoca, Romania
Luiza de Macedo Mourelle	State University of Rio de Janeiro, Brazil
Anirban Mukherjee	Indian Institute of Technology Kharagpur, India
Vittorio Murino	Istituto Italiano di Tecnologia, Italy
Marco Muselli	Consiglio Nazionale delle Ricerche, Italy
Yuichi Nakamura	Kyoto University, Japan
Michele Nappi	Università di Salerno, Italy
Fred Nicolls	University of Cape Town, South Africa
Mikael Nilsson	Lund University, Sweden
Tayo Obafemi-Ajayi	Missouri State University, USA
Hasan Ogul	Baskent University, Turkey
Il-Seok Oh	Chonbuk National University, South Korea
Simon OKeefe	University of York, UK
Luiz S. Oliveira	UFPR, Brazil
Arnau Oliver	University of Girona, Spain
Gonzalo Pajares	Universidad Complutense de Madrid, Spain
Vicente Palazón-González	Universitat Jaume I, Spain

Joao Papa	Universidade Estadual Paulista, Brazil
Danillo Pereira	University of Western São Paulo/Federal University of São Carlos, Brazil
Caroline Petitjean	LITIS EA 4108, France
Frederick Kin Hing Phoa	Academia Sinica, Taiwan
Nico Piatkowski	TU Dortmund, Germany
Luca Piras	University of Cagliari, Italy
Vincenzo Piuri	Università degli Studi di Milano, Italy
Lionel Prevost	ESIEA, College of information technologies, France
Hugo Proença	University of Beira Interior, Portugal
Rajesh Reghunadhan	Central University of Kerala, India
Bernardete M. Ribeiro	University of Coimbra, Portugal
Bryan Riley	Ohio University, USA
Juan J. Rodríguez	University of Burgos, Spain
Joseph Ronsin	INSA, IETR, France
Rosa María Valdovinos Rosas	Universidad Autonoma del Estado de Mexico, Mexico
Fernando Rubio	Universidad Complutense de Madrid, Spain
José Saavedra	Orand S.A, Chile
Robert Sabourin	Ecole de Technologie Superieure, Canada
Lorenza Saitta	Università degli Studi del Piemonte Orientale Amedeo Avogadro, Italy
Antonio-José Sánchez-Salmerón	Universitat Politecnica de Valencia, Spain
Carlo Sansone	University of Naples Federico II, Italy
K. C. Santosh	The University of South Dakota, USA
Michele Scarpiniti	Sapienza University of Rome, Italy
Paul Scheunders	University of Antwerp, Belgium
Leizer Schnitman	Universidade Federal da Bahia, Salvador, Brazil
Friedhelm Schwenker	University of Ulm, Germany
Ishwar Sethi	Oakland University, USA
Humberto Sossa	Instituto Politécnico Nacional-CIC, Mexico
Mu-Chun Su	National Central University, Taiwan
Chuan Sun	University of Central Florida, USA
Johan Suykens	KU Leuven, Belgium
Eulalia Szmidt	Systems Research Institute Polish Academy of Sciences, Poland
Alberto Taboada-Crispí	Universidad Central Marta Abreu de Las Villas, Cuba
Xiaoyang Tan	Nanjing University of Aeronautics and Astronautics, China
Oriol Ramos Terrades	Universitat Autònoma de Barcelona, Spain
Ricardo S. Torres	University of Campinas, Brazil
Genny Tortora	Università degli Studi di Salerno, Italy
Godfried Toussaint	New York University Abu Dhabi, UAE
Olgierd Unold	Wroclaw University of Technology, Poland

Ventzeslav Valev	Bulgarian Academy of Sciences, Institute of Mathematics and Informatics, Bulgaria
Ernest Valveny	Universitat Autònoma de Barcelona, Spain
Antanas Verikas	Halmstad University, Sweden
Panayiotis Vlamos	Ionian University, Greece
Asmir Vodencarevic	Siemens Healthcare GmbH, Germany
Yvon Voisin	University of Burgundy, France
Toyohide Watanabe	Nagoya Industrial Science Research Institute, Japan
Laurent Wendling	LIPADE, France
Slawomir Wierzchon	Polish Academy of Sciences, Poland
Xianghua Xie	Swansea University, UK
Jing-Hao Xue	University College London, UK
Chan-Yun Yang	National Taipei University, Taiwan
Yusuf Yaslan	Istanbul Technical University, Turkey
Olcay Yildiz	Isik University, Turkey
Nicolas Younan	Mississippi State University, USA
Slawomir Zadrozny	Polish Academy of Sciences, Poland
Danuta Zakrzewska	Lodz University of Technology, Poland
Pavel Zemcik	Brno University of Technology, Czech Republic
Bob Zhang	University of Macau, SAR China
Huiyu Zhou	Queen's University Belfast, UK
William Zhu	University of Electronic Science and Technology of China, China
Michael Zillich	Technische Universität Wien, Austria
Jacek M. Zurada	University of Louisville, USA
Reyer Zwiggelaar	Aberystwyth University, UK

Additional Reviewers

Ivan Duran-Diaz	University of Seville, Spain
Claudio Gallicchio	University of Pisa, Italy
René Grzeszick	TU Dortmund University, Germany
Antonio J. Sierra	Universidad de Sevilla, Spain
Sebastian Sudholt	Technische Universität Dortmund, Germany
Rui Zhu	University College London, UK

Invited Speakers

Isabel Trancoso	L2f INESC-ID/IST, Portugal
Vittorio Murino	Istituto Italiano di Tecnologia, Italy
Lale Akarun	Bogazici University, Turkey
Antonio Torralba	Massachusetts Institute of Technology, USA

Contents

Control Variates as a Variance Reduction Technique for Random Projections

Keegan Kang$^{(\boxtimes)}$ and Giles Hooker$^{(\boxtimes)}$

Cornell University, Ithaca, NY 14850, USA
{tk528,gjh27}@cornell.edu

Abstract. Control variates are used as a variance reduction technique in Monte Carlo integration, making use of positively correlated variables to bring about a reduction of variance for estimated data. By storing the marginal norms of our data, we can use control variates to reduce the variance of random projection estimates. We demonstrate the use of control variates in estimating the Euclidean distance and inner product between pairs of vectors, and give some insight on our control variate correction. Finally, we demonstrate our variance reduction through experiments on synthetic data and the **arcene**, **colon**, **kos**, **nips** datasets. We hope that our work provides a starting point for other control variate techniques in further random projection applications.

1 Introduction

The random projection technique is used in dimension reduction, where data in high dimensions is projected to a lower dimension using a random matrix R. One of the basic applications of this technique is to estimate the Euclidean distance and inner product between pairs of vectors.

The entries r_{ij} in the random matrix R can either be i.i.d. with mean $\mu = 0$ and second moment $\mu_2 = 1$, or correlated with each other. While it is common to have a random projection matrix R with i.i.d. entries $r_{ij} \sim N(0,1)$, speedups are achieved by having R with binary i.i.d. entries [1], or drawn from a sparse Bernoulli distribution [11]. In the above cases, the entries of the random projection matrix consists of elements $\{-1, 0, 1\}$, thus matrix multiplication is faster when compared to dense entries in $N(0,1)$.

Further speedups can be achieved by using random matrices with correlated entries, such as matrices constructed by the Lean Walsh Transform [12] to the Fast Johnson Lindenstrauss Transform (FJLT) [2] and the Subsampled Randomized Hadamard Transform (SRHT) [4]. Both these transformations make use of matrix-vector products using the Hadamard matrix, which can be computed recursively.

Consider vectors $\mathbf{x}_i \in \mathbb{R}^p$ mapped to a lower dimensional vector $\tilde{\mathbf{x}}_i \in \mathbb{R}^k$ using a random projection matrix R under the identity $\tilde{\mathbf{x}}^T = \mathbf{x}^T R$. The distance properties of these vectors $\mathbf{x}_i, \mathbf{x}_j$ are preserved in expectations in $\tilde{\mathbf{x}}_i, \tilde{\mathbf{x}}_j$. If we wanted to compute a property of $\mathbf{x}_i, \mathbf{x}_j$ given by some $f(\mathbf{x}_i, \mathbf{x}_j)$, then the goal is

© Springer International Publishing AG, part of Springer Nature 2018
M. De Marsico et al. (Eds.): ICPRAM 2017, LNCS 10857, pp. 1–20, 2018.
https://doi.org/10.1007/978-3-319-93647-5_1

to find some function $g(\cdot)$, such that $\mathbb{E}[g(\tilde{\mathbf{x}}_i, \tilde{\mathbf{x}}_j)] = f(\mathbf{x}_i, \mathbf{x}_j)$. For example, if we want an estimate of the Euclidean distance between two vectors $\mathbf{x}_i$ and $\mathbf{x}_j$, using a random projection matrix R with entries i.i.d. from $N(0,1)$, or from $\{-1,1\}$ with equal probability, then $f(a,b) = g(a,b) = \|a - b\|_2$.

Each of these resultant estimates from a chosen random matrix R have probability bounds on accuracy plus bounds on their run time, and it is up to the user to choose a random projection matrix which will suit their purposes.

In this article, we expand upon the conference proceedings *Random Projections with Control Variates* [8] in the following ways

1. We give more insight on why control variates reduce the variance of our random projection estimates.
2. We give more intuition on the control variate correction.
3. We perform more experiments on more datasets.

The control variate approach with random projections can be used with different types of different random projection matrices. This leads to a variance reduction in the estimation of Euclidean distances and inner products between pairs of vectors $\mathbf{x}_i, \mathbf{x}_j$ with a negligible extra cost in speed and storage space. Such measures of distances can be used in clustering [5,6], classification [15], and set resemblance problems [10].

We structure this article as follows: First, we express our notation differently from the ordinary random projection notation to give intuition on how we can use control variates. Next, we explain the control variate technique of variance reduction and show control variates achieve variance reduction. We next look at related work which inspired our method, before introducing the control variate corrections for Euclidean distances and inner products. Lastly, we demonstrate our method on both synthetic and experimental data and show that we can use a control variate approach together with any random projection method to gain variance reduction in our estimates.

1.1   Notation and Intuition

With classical random projections, we denote $R \in \mathbb{R}^{p \times k}$ to be a random projection matrix. We let $X \in \mathbb{R}^{n \times p}$ to be our data matrix, where each row $\mathbf{x}_i^T \in \mathbb{R}^p$ is a p dimensional observation. In most textbooks, the random projection equation is given by

$$V = \frac{1}{\sqrt{k}} X R \tag{1}$$

However, we will use

$$V = X R \tag{2}$$

without the scaling factor. The motivation is to see each element $v_{ij} \in V$ as a random variable drawn from some probability distribution.

Consider the random matrix R written as

$$R = [\mathbf{r}_1 \mid \mathbf{r}_2 \mid \ \ldots \ \mid \mathbf{r}_k] \tag{3}$$

where each $\mathbf{r}_i$ is a column vector with i.i.d. entries. Then for a *fixed* row $\mathbf{x}_i^T$, the elements $\{v_{ij}^2\}_{j=1}^k$ are drawn from the same probability distribution with mean $\|\mathbf{x}_i\|_2^2$.

By the Law of Large Numbers, we would expect that as k increases, the mean of the observations $\{v_{ij}^2\}_{j=1}^k$ would converge to the true value of $\|\mathbf{x}_i\|_2^2$.

Similarly, we have the means of $\{(v_{is}-v_{js})^2\}_{s=1}^k$ and $\{(v_{is}v_{js})\}_{s=1}^k$ converging to $\|\mathbf{x}_i - \mathbf{x}_j\|_2^2$ and $\langle \mathbf{x}_i, \mathbf{x}_j \rangle$ respectively.

For other random projection matrices R where the entries come from a different distribution, we can also find equivalent expressions of the form $\{f(\mathbf{v}_{is}, \mathbf{v}_{js})\}_{s=1}^k$ where the mean of these observations converge to either the squared Euclidean distance or inner product.

1.2   Probability Bounds on Random Projection Estimates

We give the form of the probability bounds on random projection estimates in order to show how control variates give us a tighter bound.

Suppose we look at a single row $\mathbf{v_i} \in V$, and let $S_k^{\mathrm{norm}} = \sum_{s=1}^k v_{is}^2$. By finding expressions of the form $f_1(\epsilon, k_1), f_2(\epsilon, k_2)$ where

$$\mathbb{P}\left[\frac{S_k^{\mathrm{norm}}}{k} \geq (1+\epsilon)\|\mathbf{x}\|_2^2 \right] \leq f_1(\epsilon, k_1) \tag{4}$$

$$\mathbb{P}\left[\frac{S_k^{\mathrm{norm}}}{k} \leq (1-\epsilon)\|\mathbf{x}\|_2^2 \right] \leq f_2(\epsilon, k_2) \tag{5}$$

we can then place bounds on how far our estimate of the norm is relative to our actual value since we have

$$\mathbb{P}\left[(1-\epsilon)\|\mathbf{x}\|_2^2 \leq \frac{S_k^{\mathrm{norm}}}{k} \leq (1+\epsilon)\|\mathbf{x}\|_2^2 \right] \leq 1 - f_1(\epsilon, k_1) - f_2(\epsilon, k_2) \tag{6}$$

Furthermore, computing these expressions $f_1(\epsilon, k_1), f_2(\epsilon, k_2)$ suffices to place probability bounds on our estimate of Euclidean distances and inner products. Similarly, by defining $S_k^{\mathrm{ED}} := \sum_{s=1}^k (v_{is} - v_{js})^2$ and $S_k^{\mathrm{IP}} := \sum_{s=1}^k v_{is}v_{js}$, we can thus write

$$\mathbb{P}\left[(1-\epsilon)\|\mathbf{x}_i - \mathbf{x}_j\|_2^2 \leq \frac{S_k^{\mathrm{ED}}}{k} \leq (1+\epsilon)\|\mathbf{x}_i - \mathbf{x}_j\|_2^2 \right] \leq 1 - f_1(\epsilon, k_1) - f_2(\epsilon, k_2) \tag{7}$$

$$\mathbb{P}\left[(1-\epsilon)\langle \mathbf{x}_i, \mathbf{x}_j \rangle \leq \frac{S_k^{\mathrm{IP}}}{k} \leq (1+\epsilon)\langle \mathbf{x}_i, \mathbf{x}_j \rangle_2^2 \right] \leq 1 - 2f_1(\epsilon, k_1) - 2f_2(\epsilon, k_2) \tag{8}$$

To see this, we can replace $v_{is}^2, \mathbf{x}$ in Eqs. 4 and 5 by $(v_{is} - v_{js})^2, \mathbf{x}_i - \mathbf{x}_j$ and get the bounds in (7).

We can also use the identities

$$\|\mathbf{v}_i - \mathbf{v}_j\|_2^2 = \|\mathbf{v}_i\|_2^2 + \|\mathbf{v}_j\|_2^2 - 2\langle \mathbf{v}_i, \mathbf{v}_j \rangle \tag{9}$$

$$\|\mathbf{v}_i + \mathbf{v}_j\|_2^2 = \|\mathbf{v}_i\|_2^2 + \|\mathbf{v}_j\|_2^2 + 2\langle \mathbf{v}_i, \mathbf{v}_j \rangle \tag{10}$$

and the union bound to show (8), by expressing $\langle \mathbf{v}_i, \mathbf{v}_j \rangle$ in terms of $\langle \mathbf{x}_i, \mathbf{x}_j \rangle$.

When r_{ij} are i.i.d. $N(0,1)$, we get $f_1(\epsilon, k_1) = f_2(\epsilon, k_2) = \exp\left\{-\frac{k(\epsilon^2 - \epsilon^3)}{4}\right\}$ [18]. The bounds for other random projection matrices can be found in [2,4,18].

These probability bounds are usually computed by looking at the respective second moments, and using a Chernoff bound approach.

1.3   Control Variates

Having introduced the notion of each v_{ij} as a random variable, we now look at control variates. Control variates are a technique in Monte Carlo simulation using random variables for variance reduction. A more thorough explanation can be found in Ross [17].

The method of control variates assumes we use the same random inputs to estimate $\mathbb{E}[A] = \mu_A$, for which we know B with $\mathbb{E}[B] = \mu_B$. We call B our control variate. Then to estimate $\mathbb{E}[A] = \mu_A$ from some distribution A, we can instead compute the expectation of

$$\mathbb{E}[A + c(B - \mu_B)] \quad = \quad \mathbb{E}[A] + c\mathbb{E}[B - \mu_B] \quad = \quad \mu_A \qquad (11)$$

which is an unbiased estimator of μ_A for some constant c, which is our control variate correction. This value of c which minimizes the variance is given by

$$\hat{c} = -\frac{\text{Cov}(A, B)}{\text{Var}(B)} \qquad (12)$$

and thus we write

$$\text{Var}[A + c(B - \mu_B)] = \text{Var}(A) - \frac{(\text{Cov}(A, B))^2}{\text{Var}(B)} \qquad (13)$$

Suppose we look at

$$S_k^{\text{norm}} = \sum_{j=1}^{n} v_{ij}^2 \qquad (14)$$

We can think of A being the probability distribution of the v_{is}, with the mean $\mu_A = \|\mathbf{x}_i\|_2^2$ being our target. If we found some probability distribution B of the form $f(v_{ij})$ and known mean μ', then we must have

$$S_k^{\text{cvnorm}} = \sum_{j=1}^{n} v_{ij}^2 + c(f(v_{ij}) - \mu') \qquad (15)$$

The expected value of S_k^{cvnorm} is still $\|\mathbf{x}\|_2^2$, but the second moment of S_k^{cvnorm} has to be lower (or no worse) than S_k^{norm}. To see this, recall that the variance of a distribution is given by

$$\text{Var}[A] = \mathbb{E}[A^2] - (\mathbb{E}[A])^2 \qquad (16)$$

where $\mathbb{E}[A^2]$ is the second moment. Therefore, we can take the variance of our expression as a proxy for the second moment, and by (13), have our second moment of S_k^{cvnorm} to be no greater than S_k^{norm}, since the term $\frac{(\text{Cov}(A,B))^2}{\text{Var}(B)}$ is always positive.

Therefore, we need to find some distribution B where the variables b_i are correlated with v_{ij} to get good variance reduction. If they were independent, then the numerator in our term $\frac{(\text{Cov}(A,B))^2}{\text{Var}(B)}$ becomes zero, and we do not get any variance reduction at all.

To find such a distribution B, we necessarily need to fulfill two conditions.

Condition 1: Since each realization v_{ij} is the sum of p random variables $r_{1j}, \ldots, r_{pj}$, we need to have y_i constructed from these same random variables *and* also correlated with each $x_{i1}, \ldots, x_{ip}$ in order to get a variance reduction.

Condition 2: We need to know the actual value of μ_B, the mean of B.

This seems like a chicken and egg problem since any μ_B that is related to both $x_{i\cdot}$, $r_{\cdot j}$ would be of some form of either the Euclidean distance or the inner product, both of which we want to estimate in the first place. We solve this problem by considering an expression that relates both the Euclidean distance and the inner product simultaneously.

1.4 Related Work

We draw inspiration from the works of Li et al. [9–11]. In these papers, Li *et al.* expressed the tuple (v_{is}, v_{js}) coming from a bivariate normal when the entries of the random projection matrix R is i.i.d. $N(0,1)$.

More formally, given the matrix $V = XR$ where each $r_{ij} \sim N(0,1)$, then for any two rows $\mathbf{v}_i, \mathbf{v}_j$ of V we have the tuple

$$\begin{pmatrix} v_{is} \\ v_{js} \end{pmatrix} \sim N\left(\begin{pmatrix} 0 \\ 0 \end{pmatrix}, \begin{pmatrix} m_i & a \\ a & m_j \end{pmatrix} \right) \tag{17}$$

where m_i, m_j denote the norms $\|\mathbf{x}_i\|_2^2, \|\mathbf{x}_j\|_2^2$ respectively, and a denotes the inner product $\langle \mathbf{x}_i, \mathbf{x}_j \rangle$.

Li et al. showed that if marginal information such as the actual norms $\|\mathbf{x}_i\|_2^2, \|\mathbf{x}_j\|_2^2$ were precomputed and stored, then it is possible to get a more accurate estimate of the inner product a using an asymptotic maximum likelihood estimator.

To do this, Li et al. computed the log-likelihood function after observing k such draws $\{v_{is}, v_{js}\}_{s=1}^k$ which is given by

$$l(a) \propto -\frac{1}{2}\log(m_1 m_2 - a^2) - \frac{1}{2}\frac{1}{m_1 m_2 - a^2}\sum_{s=1}^k (v_{is}^2 m_j - 2v_{is}v_{js}a + v_{js}^2 m_i) \tag{18}$$

and found the value of $\hat{a}$ which maximizes this function via root finding techniques.

Li et al. also showed that the above result also held asymptotically when the entries of the random matrix R do not come from $N(0,1)$. If they were i.i.d. from the Sparse Bernoulli distribution, then under the Central Limit Theorem, the tuple (v_{is}, v_{js}) also converges to the bivariate normal.

We will use these results below.

1.5 Our Contributions

We propose using control variates in this article to reduce the variance of the estimates of the Euclidean distances and the inner products between pairs of vectors for a choice of random projection matrix R. In particular

1. We describe the process of the control variate approach, which has the same time complexity to a non control variate approach.
2. We give the first and second moments of $A + c(B - \mu_B)$ for matrices R with i.i.d. entries, which can then be used to bound the errors in our estimates.
3. We demonstrate empirically that our control variate approach works well with current random projection methods on synthetically generated data and the `arcene`, `colon`, `kos`, `nips` datasets.

2 Process of Using Control Variates

We describe and illustrate the process of using control variates in this section.

Without loss of generality, suppose we had $\mathbf{x}_1, \mathbf{x}_2 \in \mathbb{R}^p$. Consider $\mathbf{v}$ given by $X\mathbf{r}$. For the case $p = 2$, we would have

$$V = \begin{pmatrix} v_1 \\ v_2 \end{pmatrix} = \begin{pmatrix} x_{11} & x_{12} \\ x_{21} & x_{22} \end{pmatrix} \begin{pmatrix} r_1 \\ r_2 \end{pmatrix} = X\mathbf{r} \tag{19}$$

for one column of R. We do matrix multiplication $X\mathbf{r}$ and get v_1, v_2.

In the next two sections, we will give the control variate to estimate the Euclidean distance and the inner product. We will also give the respective optimal control variate correction c, and the respective first and second moments of the expression $A + c(B - \mu_B)$. This allows us to compute a more accurate estimate for the Euclidean distance and the inner product, as well as place probability bounds on the errors of our estimates.

2.1 Control Variate for the Euclidean Distance

Suppose we computed V as above. The following theorem shows us how to estimate the Euclidean distance with our control variate.

Theorem 1. *Let one realization of $A = (v_1 - v_2)^2$, which is an estimate of our Euclidean distance. Let one realization of B to be $(v_1 - v_2)^2 + 2v_1v_2 = v_1^2 + v_2^2$ with expected value $\mu_B = \|\mathbf{x}_1\|^2 + \|\mathbf{x}_2\|_2^2$. The Euclidean distance (in expectation) between these two vectors is given by $\mathbb{E}[A + c(B - \mu_B)]$, and we can compute $c := -Cov(A, B)/Var(B)$ from our matrix V directly, using the empirical covariance $Cov(A, B)$ and empirical variance $Var(B)$.*

Proof. We have

$$\mathbb{E}[(v_1 - v_2)^2] + 2\mathbb{E}[v_1 v_2]$$

$$= \|\mathbf{x}_1 - \mathbf{x}_2\|^2 + 2\langle \mathbf{x}_1 \mathbf{x}_2 \rangle \tag{20}$$

$$= \|\mathbf{x}_1\|^2 + \|\mathbf{x}_2\|^2 - 2\langle \mathbf{x}_1, \mathbf{x}_2 \rangle + 2\langle \mathbf{x}_1, \mathbf{x}_2 \rangle \tag{21}$$

$$= \|\mathbf{x}_1\|^2 + \|\mathbf{x}_2\|^2 \tag{22}$$

We derive the following lemma to help us compute the first and second moments required.

Lemma 1. Suppose we assume that our matrix R has i.i.d. entries, where each r_{ij} has mean $\mu = 0$, second moment $\mu_2 = 1$, and fourth moment μ_4. Then

$$\mathbb{E}[A^2] = \mu_4 \sum_{j=1}^{p}(x_{1j} - x_{2j})^4 + 6 \sum_{u=1}^{p-1}\sum_{v=u+1}^{p}(x_{1u} - x_{2u})^2(x_{1v} - x_{2v})^2 \tag{23}$$

$$\mathbb{E}[B^2] = \mu_4 \sum_{j=1}^{p}(x_{1j}^4 + x_{2j}^4) + 6 \sum_{u=1}^{p-1}\sum_{v=u+1}^{p}(x_{1u}^2 x_{1v}^2 + x_{2u}^2 x_{2v}^2)$$

$$+ 4 \sum_{u=1}^{p-1}\sum_{v=u+1}^{p}(x_{1u}x_{1v}x_{2u}x_{2v}) + \mu_4 \sum_{j=1}^{p}x_{1j}^2 x_{2j}^2 + \sum_{i\neq j}x_{1i}^2 y_{2j}^p \tag{24}$$

$$\mathbb{E}[AB] = 4 \sum_{u=1}^{p-1}\sum_{v=u+1}^{p}(x_{1u} - x_{2u})(x_{1v} - x_{2v})(x_{1u}x_{1v} + x_{2u}x_{2v})$$

$$+ \mu_4 \sum_{j=1}^{p}(x_{1j} - x_{2j})^2(x_{1j}^2 + x_{2j}^2) + \sum_{i\neq j}(x_{1i} - x_{2i})^2(x_{1i}^2 + x_{2j}^2) \tag{25}$$

Proof. We repeatedly apply Lemma 2 in the Appendix.

Thus, by following Lemma 1, we are able to derive expressions for the optimal control variate correction c in our procedure as follows.

Theorem 2. *The optimal value c is given by*

$$c = -\frac{Cov(A, B)}{Var[B]} \tag{26}$$

where we have

$$Cov(A, B) = \mathbb{E}[AB - A\mu_B - B\mu_A + \mu_A\mu_B] \tag{27}$$

and

$$Var[B] = \mathbb{E}[B^2] - (\mathbb{E}[B])^2 \tag{28}$$

They expand to

$$Cov(A, B) = 4 \sum_{u=1}^{p-1} \sum_{v=u+1}^{p} (x_{1u} - x_{2u})(x_{1v} - x_{2v})(x_{1u}x_{1v} + x_{2u}x_{2v})$$

$$+ (\mu_4 - 1) \sum_{j=1}^{p} (x_{1j} - x_{2j})^2 (x_{1j}^2 + x_{2j}^2) \tag{29}$$

and

$$Var[B] = (\mu_4 - 1) \sum_{j=1}^{p} (x_{1j}^4 + x_{2j}^4) + 4 \sum_{u=1}^{p-1} \sum_{v=u+1}^{p} (x_{1u}^2 x_{1v}^2$$

$$+ x_{2u}^2 x_{2v}^2) + 4 \sum_{u=1}^{p-1} \sum_{v=u+1}^{p} x_{1u}x_{1v}x_{2u}x_{2v} + (\mu_4 - 2) \sum_{j=1}^{p} x_{1j}^2 x_{2j}^2 - \sum_{i \neq j} x_{1i}^2 x_{2j}^2 \tag{30}$$

We are also able to derive the first and second moments of $A + c(B - \mu_B)$ for Euclidean distances.

Theorem 3. *The first and second moments are*

$$\mathbb{E}[A + c(B - \mu_B)] = \mathbb{E}[A] + c\mathbb{E}[B - \mu_B] = 0 \tag{31}$$

and

$$\mathbb{E}[(A + c(B - \mu_B))^2] = \mathbb{E}[A^2 + 2cAB - 2c\mu_B A + c^2 B^2 - 2c^2 \mu_B B + c^2 \mu_B^2] \tag{32}$$

where we substitute in the values of $\mathbb{E}[A^2], \mathbb{E}[AB], \mathbb{E}[B^2]$ *from Lemma 1.*

These first and second moments could be used to get tighter (and exact) bounds of the form in (4) and (5) by using a Chernoff bound type strategy.

2.2   Control Variate for the Inner Product

Suppose we computed V as above. The following theorem shows us how to estimate the inner product with our control variate.

Theorem 4. *Let one realization of* $A = v_1 v_2$, *which is an estimate of our inner product. Let one realization of* B *to be* $(v_1 - v_2)^2 + 2v_1 v_2 = v_1^2 + v_2^2$ *with expected value* $\mu_B = \|\mathbf{x}_1\|^2 + \|\mathbf{x}_2\|_2^2$. *The inner product between these two vectors is given by* $\mathbb{E}[A + c(B - \mu_B)]$, *and we can compute* $c := -Cov(A, B)/Var(B)$ *from our matrix* V *directly, using the empirical covariance* $Cov(A, B)$ *and empirical variance* $Var(B)$.

The optimal control variate c in this procedure is given by the next theorem.

Theorem 5. *The optimal value of c is given by*

$$c = -\frac{Cov(A, B)}{Var[B]} \tag{33}$$

where

$$Cov(A, B) = \mathbb{E}[AB - A\mu_B - B\mu_A + \mu_A\mu_B]$$

$$= (\mu_4 - 1)\sum_{j=1}^{p} x_{1j}x_{2j}(x_{1j}^2 + x_{2j}^2) + \sum_{i \neq j} x_{1i}x_{2j}(x_{1i}x_{1j} + x_{2i}x_{2j}) \tag{34}$$

and the value of Var[B] taken from the result in Theorem 2.

2.3   The Optimal Control Variate Correction c

While we have computed an expression c in terms of the first and second moments of our distributions, they are not at all intuitive from first sight. Therefore, we consider what the optimal value of c would be if the random matrix R had i.i.d. entries $r_{ij} \sim N(0, 1)$. Thus, we take a second look at the bivariate normal distribution in (17).

Theorem 6. *For $r_{ij} \sim N(0, 1)$, and $V = XR$, the optimal control variate correction c_{ED} for the Euclidean distance is given by*

$$c_{ED} = -\frac{(m_i - a)^2 + (m_j - a)^2}{(m_i^2 + m_j^2 + 2a^2)} \tag{35}$$

We use m_i, m_j to denote the norms $\|\mathbf{x}_i\|_2^2, \|\mathbf{x}_j\|_2^2$ respectively, and a to be $\langle \mathbf{x}_i, \mathbf{x}_j \rangle$ as in (17).

Proof. We can write the control variate correction c for the Euclidean distance as

$$c_{ED} = -\frac{Cov(v_i^2 + v_j^2, v_i^2 + v_j^2 - 2v_iv_j)}{Var\left(v_i^2 + v_j^2\right)} \tag{36}$$

$$= -\frac{Cov(\mathbf{v}_i^T\mathbf{v}_j, \mathbf{v}_i^T H\mathbf{v}_j)}{Var\left(\mathbf{v}_i^T\mathbf{v}_j\right)} \tag{37}$$

where $H = \begin{pmatrix} 1 & -1 \\ -1 & 1 \end{pmatrix}$. Next, expanding the numerator gives

$$Cov(\mathbf{v}_i^T\mathbf{v}_j, \mathbf{v}_i^T H\mathbf{v}_j) = \mathbb{E}[\mathbf{v}_i^T\mathbf{v}_j\mathbf{v}_i^T H\mathbf{v}_j] - \mathbb{E}\left[\mathbf{v}_i^T\mathbf{v}_j\right]\mathbb{E}\left[\mathbf{v}_i^T H\mathbf{v}_j\right] \tag{38}$$

For $(v_i, v_j) \sim N(0, \Sigma)$, we have the identities

$$\mathbb{E}\left[\mathbf{v}_i^T\mathbf{v}_j\mathbf{v}_i^T H\mathbf{v}_j\right] = Tr(\Sigma(H + H^T)\Sigma) + Tr(\Sigma)Tr(H\Sigma) \tag{39}$$

$$= 2(m_i - a)^2 + 2(m_j - a)^2 + (m_i + m_j)(m_i + m_j - 2a) \tag{40}$$

$$\mathbb{E}[\mathbf{v}_i^T\mathbf{v}_j] = m_i + m_i \tag{41}$$

$$\mathbb{E}[\mathbf{v}_i^T H\mathbf{v}_j] = m_j + m_j - 2a \tag{42}$$

and therefore, we have

$$\text{Cov}(\mathbf{v}_i^T \mathbf{v}_j, \mathbf{v}_i^T H \mathbf{v}_j) = 2(m_i - a)^2 + 2(m_j - a)^2 \qquad (43)$$

The denominator expands to be

$$\text{Var}\left(\mathbf{v}_i^T \mathbf{v}_j\right) = \text{Tr}\left(\Sigma(2I)\Sigma\right) \qquad (44)$$

$$= 2(m_i^2 + m_j^2 + 2a^2) \qquad (45)$$

Simplifying, we get:

$$c_{\text{ED}} = -\frac{(m_i - a)^2 + (m_j - a)^2}{(m_i^2 + m_j^2 + 2a^2)} \qquad (46)$$

Theorem 7. *For $r_{ij} \sim N(0,1)$, and $V = XR$, the optimal control variate correction c_{IP} for the inner product is given by*

$$c_{IP} = -\frac{m_i a + m_j a}{m_i^2 + m_j^2 + 2a^2} \qquad (47)$$

Proof. Analogous to the proof of Theorem 6, we express

$$\text{Cov}(v_i^2 + v_j^2, v_i v_j) = \text{Cov}\left(\mathbf{v}_i^2 \mathbf{v}_j, \mathbf{v}_i^T H \mathbf{v}_j\right) \qquad (48)$$

where $H = \frac{1}{2}\begin{pmatrix} 0 & 1 \\ 1 & 0 \end{pmatrix}$. Therefore, we similarly compute

$$\mathbb{E}\left[\mathbf{v}_i^T \mathbf{v}_j \mathbf{v}_i^T H \mathbf{v}_j\right] = \text{Tr}(\Sigma(H + H^T)\Sigma) + \text{Tr}(\Sigma)\text{Tr}(H\Sigma) \qquad (49)$$

$$= 2m_i a + 2m_j a + (m_i + m_j)(a) \qquad (50)$$

$$\mathbb{E}[\mathbf{v}_i^T \mathbf{v}_j] = m_i + m_j \qquad (51)$$

$$\mathbb{E}[\mathbf{v}_i^T H \mathbf{v}_j] = a \qquad (52)$$

which results in

$$c_{\text{IP}} = \frac{m_i a + m_j a}{m_i^2 + m_j^2 + 2a^2} \qquad (53)$$

Without loss of generality, we assume that our data is normalized such that $m_i = m_j = 1$. In this case, we can compute the variance reduction for our Euclidean distances and inner products respectively.

Theorem 8. *Given $r_{ij} \sim N(0,1)$ and $V = XR$, then for any pair $\mathbf{x}_i, \mathbf{x}_j$*

1. *The variance of the estimate of the Euclidean distance between the pair is given by*

$$\sigma_{ED} = 8(1 - a)^2 \qquad (54)$$

2. *The variance of the estimate of the Euclidean distance with the control variate correction between the pair is given by*

$$\sigma_{EDCV} = 8(1 - a)^2 - \frac{4(1 - a)^4}{(1 + a^2)} \qquad (55)$$

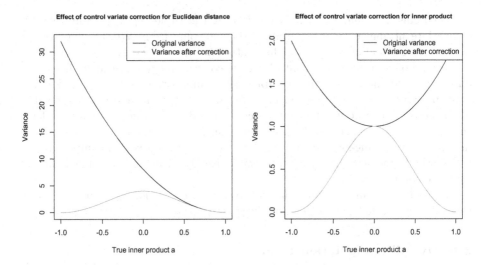

Fig. 1. Effects of control variate correction on estimates.

3. *The variance of the estimate of the inner product between the pair is given by*

$$\sigma_{IP} = 1 + a^2 \tag{56}$$

4. *The variance of the estimate of the inner product with the control variate correction between the pair is given by*

$$\sigma_{IPCV} = 1 + a^2 - \frac{4a^2}{1 + a^2} \tag{57}$$

Proof. This follows from direct substitution of the optimal control variate corrections in Theorems 6 and 7 into (13).

Theorem 8 allows us to analyze the effect of our control variate correction, given in Fig. 1.

Recall that when observations are normalized, we have the Euclidean distance between any pair of vectors being in the range $[0, 2]$, and the inner product between any pair of vectors being in the range $[0, 1]$.

We can now analyze the effect of our control variate correction on Euclidean distance. For vectors $\mathbf{x}_i, \mathbf{x}_j$ in the same direction (inner product close to 1), we do not get much variance reduction in our estimate of Euclidean distance. Conversely, if the vectors were in opposite directions, then we would get a reasonable variance reduction in the estimates of their Euclidean distance.

Similarly, we analyze the effect of our control variate correction on the inner product. If the vectors $\mathbf{x}_i, \mathbf{x}_j$ are orthogonal to each other (inner product near 0), then we do not get much variance reduction from our control variate correction. Conversely, if the vectors share the same or opposite directions, then we would get a reasonable variance reduction in the estimates of their inner product.

Theorems 6, 7, and Li *et al.*'s result that the tuple (v_{is}, v_{js}) converges to a bivariate normal even if r_{ij} from R do not come from $N(0,1)$ suggests an alternative method of computing the control variate correction.

Instead of computing the control variate correction c_{ED} empirically from our data, we could choose to either compute the vanilla estimate $\hat{a} = v_{is}v_{js}$ for the inner product, or $\hat{a}$ using Li's method. We can then substitute $\hat{a}$ into the results of Theorem 6 to compute the optimal control variate correction, using the fact that we get convergence to a bivariate normal when the number of observations increase. In fact, since we are storing the marginal norms, we can compute $\hat{a}$ via Li's method, and use this to compute c_{ED} directly since this does not increase the time complexity.

Similarly, we could compute the control variate correction c_{IP}, but only using the ordinary estimate $\hat{a} = v_{is}v_{js}$ for the inner product.

2.4   Overall Computational Time

Constructing the matrix $V = X_{n \times p} R_{p \times k}$ takes $O(npk)$ time, and computing the pairwise Euclidean distances (or inner products) takes an additional $O(n^2 k)$ of time.

If we want to use control variates, we either need to compute the empirical covariance between all pairs A and B and the variance of B, or compute an estimate of the inner product a to put into our control variate correction. Both these options take an additional $O(nk)$ time for all pairwise computations. We also need to compute and store the norm of each vector, which takes $O(np)$ time.

Thus, the overall computational time is given by $O(npk + n^2 k + n(k+p)) = O(npk + n^2 k)$.

3   Our Experiments

Throughout our experiments, we use five different types of random projection matrices as shown in Table 1 (also used in [8]). We pick these five types of random projection matrices as they are commonly used random projection matrices.

We use $N(0,1)$ to denote the Normal distribution with mean $\mu = 0$ and $\sigma^2 = 1$. We denote $(\mathbf{1})_p$ to be the length p vector with all entries being 1, and $(\mathbf{0})_p$ to be the length p vector with all entries being 0. We denote the baseline estimates to be the respective estimates given by using the type of random projection matrix R_i.

We run our simulations for 10000 iterations for every experiment.

3.1   Generating Vectors from Synthetic Data

We first perform our experiments on a wide range of synthetic data. We look at normalized pairs of vectors $\mathbf{x}_1$, $\mathbf{x}_2 \in \mathbb{R}^{5000}$ generated from the following distributions in Table 2 (also used in [8]). In short, we look at data that can be

Table 1. Random projection matrices.

R	Type
R_1	Entries i.i.d. from $N(0,1)$
R_2	Entries i.i.d. from $\{-1,1\}$ with equal probability
R_3	Entries i.i.d. from $\{-\sqrt{p}, 0, \sqrt{p}\}$ with probabilities $\{\frac{1}{2p}, 1 - \frac{1}{p}, \frac{1}{2p}\}$ for $p = 5$
R_4	Entries i.i.d. from $\{-\sqrt{p}, 0, \sqrt{p}\}$ with probabilities $\{\frac{1}{2p}, 1 - \frac{1}{p}, \frac{1}{2p}\}$ for $p = 10$
R_5	Constructed using the Subsampled Randomized Hadamard Transform (SRHT)

Table 2. Generated data $\mathbf{x}_1$, $\mathbf{x}_2$.

Pairs	$\mathbf{x}_1$	$\mathbf{x}_2$
Pair 1	Entries i.i.d. from $N(0,1)$	Entries i.i.d. from $N(0,1)$
Pair 2	Entries i.i.d. from standard Cauchy	Entries i.i.d. from standard Cauchy
Pair 3	Entries i.i.d. from Bernoulli(0.05)	Entries i.i.d. from Bernoulli(0.05)
Pair 4	Vector $[(\mathbf{1})_{p/2}, (\mathbf{0})_{p/2}]$	Vector $[(\mathbf{0})_{p/2}, (\mathbf{1})_{p/2}]$

Normal, heavy tailed (Cauchy), sparse (Bernoulli), and an adversarial scenario where the inner product is zero.

We first compare the relative bias of the estimates of Euclidean distance using our control variate approach against the baseline estimates for each projection R_i for a sanity check. Plots can be seen in Fig. 2, and the relative bias goes to zero as expected.

We then look at the plots of the ratio ρ defined by

$$\rho = \frac{\text{Variance using control variate with } R_i}{\text{Variance using baseline with } R_i} \tag{58}$$

in Fig. 3 for the Euclidean distance (also used in [8]). ρ is a measure of the reduction in variance using our control variate approach with the matrix R_i rather than just using R_i alone. For this ratio, a fraction less than 1 means our control variate approach performs better than the baseline.

For all pairs $\mathbf{x}_i, \mathbf{x}_j$ except Cauchy, the reduction of variance of the estimates of the Euclidean distance using different R_is with our control variate approach converge quickly to around the same ratio. However, when data is heavy tailed, the choice of random projection matrix R_i with a control variate approach affects the reduction of variance in the estimates of the Euclidean distance, and sparse matrices R_i have a greater variance reduction for the estimates of the Euclidean distance.

We next look at the estimates of the inner product. In our experiments, we use Li $et\ al.$'s method as the baseline for computing the estimates of the inner product. Our rationale for doing this is that both Li's method and our method

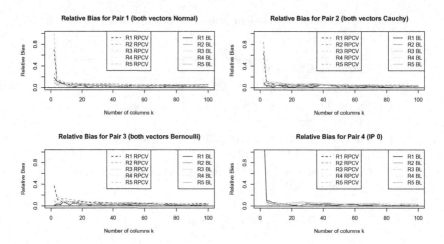

Fig. 2. Plots of relative bias of estimates of Euclidean distances against number of columns in R_i for each pair of vectors.

stores the marginal norms of X, thus we should compare our method with Li's method for a fair comparison. The ratio of variance reduction is shown in Fig. 4 (also used in [8]).

As the number of columns k of the random projection matrix R increases, the variance reduction in our estimate of the inner product decreases, but then increases again up to a ratio just below 1. Since Li's method uses an asymptotic maximum likelihood estimate of the inner product, then as the number of columns of R increases, the estimate of the inner product would be more accurate.

Thus, it is reasonable to use a control variate approach for Euclidean distances, and Li's method for inner products.

3.2   Experiments with Real Data

We now demonstrate our control variate approach on four datasets, the `arcene` dataset, `colon` dataset, `kos` dataset, and the `NIPS` dataset. We select these datasets since they have different characteristics (sparse/dense, variance explained/not explained in few principal components). In short

1. the `arcene` dataset [7] is an example of a dense dataset consisting of $n = 900$ observations with $p = 10000$ features. Most of the variance in this dataset is explained by the first 500 eigenvectors.
2. the `colon` dataset [3] is an example of a dense dataset consisting of $n = 62$ gene expression levels with 2000 features. Most of the variance in this dataset is explained by a few eigenvectors.
3. the `kos` dataset [13] is an example of a sparse dataset consisting of $n = 3430$ documents and $p = 6906$ words from the KOS blog entries. Most of the variance in this dataset is explained by about a third of the eigenvectors.

4. the **nips** dataset [16] is an example of a sparse dataset consisting of $n = 5812$ observations (conference papers) in this dataset, and $p = 11463$ words. Most of the variance in this dataset is explained by slightly less than half of the eigenvectors.

We normalize each dataset such that every observation $\|\mathbf{x}_i\|_2^2 = 1$.

For each dataset, we consider the pairwise Euclidean distances of all observations $\{\mathbf{x}_i, \mathbf{x}_j\}$, $\forall\ i \neq j$, and compute the estimates of the Euclidean distance with a control variate approach of the pairs $\{\mathbf{x}_i, \mathbf{x}_j\}$ which give the 20th, 30th, ..., 90th percentile of Euclidean distances.

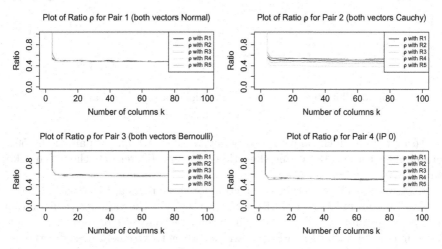

Fig. 3. Plots of ρ for Euclidean distances against number of columns in R_i for each pair of vectors.

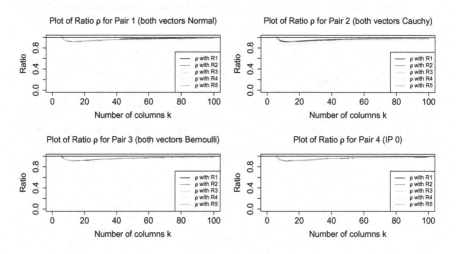

Fig. 4. Plots of ρ for inner product against number of columns in R_i for each pair of vectors.

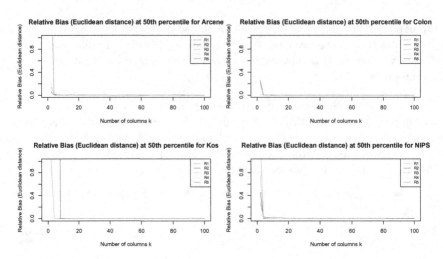

Fig. 5. Plots of relative bias in Euclidean distance for real data.

We first do a quick sanity check in Fig. 5. Here, we pick a pair in the 50th percentile for these datasets and show that for every different R_i, the bias quickly converges to zero.

Next, we look at the variance reduction for these pairs in Fig. 6 with different types of random projection matrices R_i. We see that the variance reduction for the R_is are around the same range. Since the bias converges to zero, this implies that our control variates work. i.e., we do not get extremely biased estimates with lower variance.

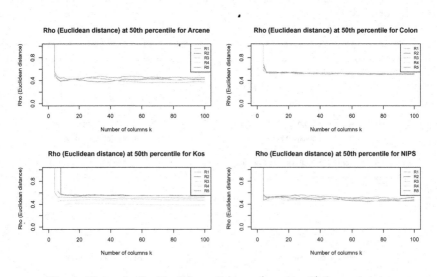

Fig. 6. Plots of ρ for Euclidean distance (varying R) for real data.

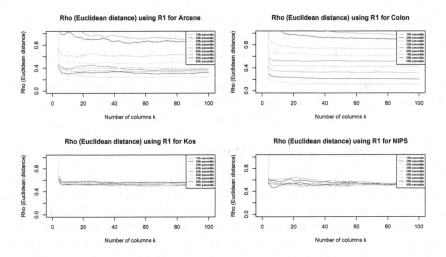

Fig. 7. Plots of ρ for Euclidean distance (varying percentile) for real data.

We now look at what happens at different percentile pairs. Since the random projection matrices have a similar pattern in Fig. 6, we will only take a look at varying pairs for the random projection matrix R_1.

Figure 7 thus shows the ratio ρ of variance reduction from the 10th percentile to the 90th percentile. Note that for dense datasets (`arcene`, `colon`), we can see a substantial percentage increase in variance reduction as the percentiles increase, but not as much for sparse datasets (`kos`, `NIPS`).

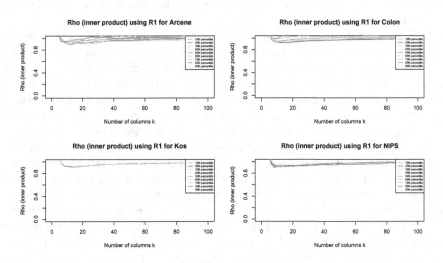

Fig. 8. Plots of ρ for inner product (varying percentile) for real data.

Finally, we take a look at the inner product estimates. We do not get good variance reduction results, when we used Li's method as a baseline.

Figure 8 shows the plots of ρ with the random projection matrix R_1 for varying percentiles. The same pattern holds for different types of matrices R_2 to R_4, and is similar to what we saw in synthetic data. While there is some small variance reduction, the ratio ρ quickly converges to a value near 1.

This matches what we see in our synthetic data.

4   Conclusion and Future Work

We have shown that which works well in conjunction with different random projection matrices to reduce the variance of the estimates of the Euclidean distance and inner products on different types of vectors $\mathbf{x}_i$, $\mathbf{x}_j$. This allows for more accurate estimates of the Euclidean distance. As the Euclidean distance between two vectors increases, we expect greater variance reduction. In essence, we have shown that it is possible to juxtapose statistical variance reduction methods with random projections to give better results.

While a control variate approach gives a variance reduction for the estimates of the inner products, the ratio of variance reduction becomes minimal as the number of columns increases when compared to Li's method. This is not surprising since Li's method for estimating the inner products is an asymptotic maximum likelihood estimator, and is extremely accurate as the number of columns of the random projection matrix increases.

Although a control variate approach requires storing marginal norms and computing the covariance between two p dimensional vectors, the cost of doing so is negligible when compared to matrix multiplication. Furthermore, the computation of marginal norms is unnecessary when the data is already normalized to have norm of 1.

In fact, a control variate approach can be seen as a method that nicely complements Li's method since both methods require storing marginal norms. This approach substantially reduces the errors of the estimates of the Euclidean distance, while Li's method substantially reduces the errors of the estimates of the inner product. The estimate of the inner product given by Li's method can even be used in computing the control variate correction c_{ED}, instead of evaluating the empirical value of c_{ED} directly, which is less costly.

We note that different applications may require a certain type of random projection matrix. Thus if we want to reduce the errors in our estimates, we cannot just switch to a different random projection matrix where the entries allow us to place sharper probability bounds on our errors. If we want data to be invariant under rotations, then a Normal random projection matrix would be best suited [14]. If we wanted to desparsify data, then a random projection matrix with i.i.d. entries from $\{-\sqrt{p}, 0, \sqrt{p}\}$, p small might be preferred [1]. If we are focused on speed and quick information retrieval, then very sparse random projections [11] or random projection matrices formed by the SHRT [4] would be more preferable. A control variate approach allows us to reduce the error in all these estimates.

We believe our work can be extended by looking at a control variate approach for other types of random projection matrices, such as sign random projections.

We further hope that this control variate approach can be adopted to current algorithms using random projections which require the computation of Euclidean distances (or inner products).

Lastly, Fig. 1 suggests that we could adopt a multiple control variate approach using several dominant eigenvectors of the data as control variates, since these eigenvectors would point in the general direction of the variance, and we are currently exploring this idea.

Appendix

We use the following lemma for ease of computation of first and second moments.

Lemma 2. Suppose we have a sequence of terms $\{t_i\}_{i=1}^{p} = \{a_i r_i\}_{i=1}^{p}$ for $\mathbf{a} = (a_1, a_2, \ldots, a_p)$, $\{s_i\}_{i=1}^{p} = \{b_i r_i\}_{i=1}^{p}$ for $\mathbf{b} = (b_1, b_2, \ldots, b_p)$ and r_i i.i.d. random variables with $\mathbb{E}[r_i] = 0$, $\mathbb{E}[r_i^2] = 1$ and finite third, and fourth moments, denoted by μ_3, μ_4 respectively. Then:

$$\mathbb{E}\left[\left(\sum_{i=1}^{p} t_i\right)^2\right] = \sum_{i=1}^{p} a_i^2 = \|\mathbf{a}\|_2^2 \tag{59}$$

$$\mathbb{E}\left[\left(\sum_{i=1}^{p} t_i\right)^4\right] = \mu_4 \sum_{i=1}^{p} a_i^4 + 6 \sum_{u=1}^{p-1} \sum_{v=u+1}^{p} a_u^2 a_v^2 \tag{60}$$

$$\mathbb{E}\left[\left(\sum_{i=1}^{p} s_i\right)\left(\sum_{i=1}^{p} t_i\right)\right] = \sum_{i=1}^{p} a_i b_i = \langle \mathbf{a}, \mathbf{b} \rangle \tag{61}$$

$$\mathbb{E}\left[\left(\sum_{i=1}^{p} s_i\right)^2\left(\sum_{i=1}^{p} t_i\right)^2\right] = \sum_{i=1}^{p} a_i^2 b_i^2 + \sum_{i \neq j} a_i^2 b_j^2 + 4 \sum_{u=1}^{p-1} \sum_{v=u+1}^{p} a_u b_u a_v b_v \tag{62}$$

The motivation for this lemma is that we do expansion of terms of the above four forms to prove our theorems.

References

1. Achlioptas, D.: Database-friendly random projections: Johnson-Lindenstrauss with binary coins. J. Comput. Syst. Sci. **66**(4), 671–687 (2003). https://doi.org/10.1016/S0022-0000(03)00025-4
2. Ailon, N., Chazelle, B.: The fast Johnson-Lindenstrauss transform and approximate nearest neighbors. SIAM J. Comput. **39**(1), 302–322 (2009). https://doi.org/10.1137/060673096
3. Alon, U., Barkai, N., Notterman, D., Gish, K., Ybarra, S., Mack, D., Levine, A.: Broad patterns of gene expression revealed by clustering analysis of tumor and normal colon tissues probed by oligonucleotide arrays. Proc. Nat. Acad. Sci. **96**(12), 6745–6750 (1999)

4. Boutsidis, C., Gittens, A.: Improved matrix algorithms via the subsampled randomized hadamard transform. CoRR abs/1204.0062 (2012). http://arxiv.org/abs/1204.0062

5. Boutsidis, C., Zouzias, A., Drineas, P.: Random projections for k-means clustering. In: Lafferty, J.D., Williams, C.K.I., Shawe-Taylor, J., Zemel, R.S., Culotta, A. (eds.) Advances in Neural Information Processing Systems 23, pp. 298–306. Curran Associates, Inc. (2010). http://papers.nips.cc/paper/3901-random-projections-for-k-means-clustering.pdf

6. Fern, X.Z., Brodley, C.E.: Random projection for high dimensional data clustering: a cluster ensemble approach, pp. 186–193 (2003)

7. Guyon, I., Gunn, S., Ben-Hur, A., Dror, G.: Result analysis of the NIPS 2003 feature selection challenge. In: Saul, L.K., Weiss, Y., Bottou, L. (eds.) Advances in Neural Information Processing Systems 17, pp. 545–552. MIT Press (2005). http://papers.nips.cc/paper/2728-result-analysis-of-the-nips-2003-feature-selection-challenge.pdf

8. Kang, K., Hooker, G.: Random projections with control variates. In: Proceedings of ICPRAM, February 2016

9. Li, P., Church, K.W.: A sketch algorithm for estimating two-way and multi-way associations. Comput. Linguist. **33**(3), 305–354 (2007). https://doi.org/10.1162/coli.2007.33.3.305

10. Li, P., Hastie, T.J., Church, K.W.: Improving random projections using marginal information. In: Lugosi, G., Simon, H.U. (eds.) COLT 2006. LNCS (LNAI), vol. 4005, pp. 635–649. Springer, Heidelberg (2006). https://doi.org/10.1007/11776420_46. http://dblp.uni-trier.de/db/conf/colt/colt2006.html#LiHC06

11. Li, P., Hastie, T.J., Church, K.W.: Very sparse random projections. In: Proceedings of the 12th ACM SIGKDD International Conference on Knowledge Discovery and Data Mining, KDD 2006, pp. 287–296. ACM, New York (2006). https://doi.org/10.1145/1150402.1150436

12. Liberty, E., Ailon, N., Singer, A.: Dense fast random projections and lean walsh transforms. In: Goel, A., Jansen, K., Rolim, J.D.P., Rubinfeld, R. (eds.) APPROX/RANDOM 2008. LNCS, vol. 5171, pp. 512–522. Springer, Heidelberg (2008). https://doi.org/10.1007/978-3-540-85363-3_40. http://dblp.uni-trier.de/db/conf/approx/approx2008.html#LibertyAS08

13. Lichman, M.: UCI machine learning repository (2013). http://archive.ics.uci.edu/ml

14. Mardia, K.V., Kent, J.T., Bibby, J.M.: Multivariate Analysis. Academic Press, Cambridge (1979)

15. Paul, S., Boutsidis, C., Magdon-Ismail, M., Drineas, P.: Random projections for support vector machines. CoRR abs/1211.6085 (2012). http://arxiv.org/abs/1211.6085

16. Perrone, V., Jenkins, P.A., Spano, D., Teh, Y.W.: Poisson random fields for dynamic feature models (2016). arXiv e-prints: arXiv:1611.07460

17. Ross, S.M.: Simulation, 4th edn. Academic Press Inc., Orlando (2006)

18. Vempala, S.S.: The Random Projection Method. DIMACS Series in Discrete Mathematics and Theoretical Computer Science, vol. 65, pp. 101–105. American Mathematical Society, Providence (2004). http://opac.inria.fr/record=b1101689

Graph Classification with Mapping Distance Graph Kernels

Tetsuya Kataoka, Eimi Shiotsuki, and Akihiro Inokuchi[✉]

School of Science and Technology, Kwansei Gakuin University,
2-1 Gakuen, Sanda, Hyogo, Japan
{TKataoka,inokuchi}@kwansei.ac.jp

Abstract. Graph mining is of great interest because knowledge discovery from structured data can be applied to real-world datasets. Recent improvements in system throughput have led to the need for the analysis of a large number of graphs using methods such as graph classification, the objective of which is to classify graphs of similar structures into the same class. Existing methods for representing graphs can result in difficulties such as the loss of structural information, which can be overcome using specifically designed graph kernels. In this paper, we propose two novel graph kernels, mapping distance kernel with stars (MDKS) and mapping distance kernel with vectors (MDKV), to classify labeled graphs more accurately than existing methods. The MDKS is based on the graph edit distance using star structures, and the MDKV is based on the graph edit distance using the linear sum assignment problem and graph relabeling. Because MDKS uses only small local structures that consist of adjacent vertices of each vertex in graphs, it is not substantially superior to conventional graph kernels. However, the MDKV uses local structures that consist of vertices that are reachable within a small number of steps from each vertex in graphs and, unlike existing methods, do not require isomorphism matching. In addition, we investigate a framework for computing the approximate graph edit distance between two graphs using the linear sum assignment problem (LSAP), because the proposed graph kernels are related to methods for computing the graph edit distance using LSAP.

Keywords: Graph classification · Graph kernel
Graph edit distance · Graph relabeling

1 Introduction

A graph is one of the most natural means of representing structured data. For instance, a chemical compound can be represented as a graph in which each vertex corresponds to an atom, each edge corresponds to a bond between two atoms, and the label of each vertex corresponds to the atom type. With recent improvements in system throughput, the need to analyze large numbers of graphs has arisen, and the topic of graph mining has received considerable interest

© Springer International Publishing AG, part of Springer Nature 2018
M. De Marsico et al. (Eds.): ICPRAM 2017, LNCS 10857, pp. 21–44, 2018.
https://doi.org/10.1007/978-3-319-93647-5_2

because the knowledge present in structured data can be applied to various real-world datasets. For example, in cheminformatics, certain properties of chemical compounds (e.g., mutagenicity or toxicity) can be identified by analyzing their structural information, and in bioinformatics, the prediction of protein–protein interactions is beneficial for drug discovery.

When analyzing datasets of graphs, one of the most critical measures is the dissimilarity (or similarity) among the graphs. Two representative frameworks for measuring the dissimilarity (or similarity) are based on the graph edit distance [21] and graph relabeling [13]. The graph edit distance between graphs g_i and g_j is defined as the minimum length of the sequence of edit operations required to transform g_i into g_j, where one edit operation includes the insertion or deletion of a vertex/edge or substitution of a vertex/edge label. The problem of obtaining the exact graph edit distance between graphs is known to be NP-hard. Graph relabeling iteratively relabels vertex labels in graphs using the adjacent vertices of each vertex, and then measures the similarity between sets of vertices in the graphs using the Jaccard index.

At the International Conference on Pattern Recognition Applications and Methods in 2017, we proposed two graph kernels that are more accurate than existing graph kernels by incorporating characteristics of the aforementioned frameworks [14]. The proposed graph kernels are called the mapping distance kernel with stars (MDKS) and mapping distance kernel with vectors (MDKV). The former kernel is based on a method for approximately measuring the graph edit distance between two graphs. Its computational complexity is $O(v^3)$ for two graphs, where v is the maximum number of vertices in the graphs. It sums the edit distances among star structures of height one obtained from the graphs. When the height of the star structures is increased to avoid loss of structural information, the number of vertices in each star structure exponentially increases, which prevents the efficient computation of this graph kernel. To overcome this difficulty, in the second proposed graph kernel, each of the star structures of height higher than one is represented as a vector, and the graph kernel is computed by summing the Euclidean distances between these vectors. The graph kernel between two graphs is computed in $O(h(v^3 + |\Sigma|v^2))$, where Σ is a set of vertex labels, and graphs are iteratively relabeled h times.

After proposing the graph kernels at the conference, we found that the proposed graph kernels are related to methods for computing the graph edit distance approximately using the linear sum assignment problem (LSAP) [24]. To measure the approximate graph edit distance or compute graph kernels efficiently, various substructures in graphs such as paths, walks, stars, trees, and limited-size graphs are utilized. This paper also surveys several methods for computing the approximate graph edit distance between two graphs using LSAP [3,9,23,28] and several conventional graph kernels using the substructures [1,2,5,8,12,18,22,26].

The remainder of this paper is organized as follows: In Sect. 2, we formalize the graph classification problem that we consider in this paper and explain the kernel function used in SVMs. In Sect. 3, we propose the MDKS and MDKV after we explain two graph kernel frameworks. In Sect. 4, we verify the computational

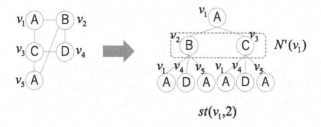

$st(v_1,2)$

Fig. 1. Subtree for v in a graph ($h = 2$).

efficiency of the proposed graph kernels on an artificially generated dataset, and compare the proposed graph kernels with conventional graph kernels in terms of classification accuracy using real-world datasets. In Sect. 5, we discuss the framework based on the graph edit distance and various graph kernels. Finally, we conclude the paper in Sect. 6.

2   Preliminaries

In this paper, we consider the graph classification problem. First, we define some terminology used for solving the problem. An undirected graph is represented as $g = (V, E, \Sigma, \ell)$, where V is a set of vertices, $E \subseteq V \times V$ is a set of edges, $\Sigma = \{\sigma_1, \sigma_2, \cdots, \sigma_\Sigma\}$ is a set of vertex labels, and $\ell : V \to \Sigma$ is a function that assigns a label to each vertex in the graph. Additionally, the set of vertices in graph g is represented as $V(g)$. Although we assume that only the vertices in the graphs have labels, the methods used in this paper can be applied to graphs where both the vertices and edges have labels [11]. The vertices adjacent to vertex v are represented as $N(v) = \{u \mid (v, u) \in E\}$. Furthermore, $L(N(v)) = \{\ell(u) \mid u \in N(v)\}$ is a multiset of labels adjacent to v.

A sequence of vertices from v to u is called a path, and its step refers to the number of edges on that path. A path is called simple if and only if the path does not have repeating vertices. Paths in this paper are not always simple. Given $v \in V(g)$, $st(v, h)$ is a subtree of height h, where v is the root and u is the child of w if u and w are adjacent in g, where the height of the subtree is the length of a path from the rooted vertex to a leaf vertex, and $N'(v')$ is a set of children of v' in $st(v, h)$. Figure 1 shows an example of a subtree of height two in a graph. As shown in Fig. 1, when a vertex v_j belongs to $N'(v_1)$ and $h > 1$, v_1 also belongs to $N'(v_j)$; that is, v' is a grandchild of v' in $st(v, h)$.

The graph classification problem is defined as follows. Given a set of n training examples $D = \{(g_i, y_i)\}$ ($i = 1, 2, \cdots, n$), where each example is a pair that consists of a labeled graph g_i and the class $y_i \in \{+1, -1\}$ to which it belongs, the objective is to learn a function f that correctly classifies the classes of the test examples.

We can classify graphs using an SVM and Gaussian kernel. Given two examples $\boldsymbol{x}_i$ and $\boldsymbol{x}_j$ as feature vectors, the Gaussian kernel function $k(\boldsymbol{x}_i, \boldsymbol{x}_j)$ is defined as

$$k(\boldsymbol{x}_i, \boldsymbol{x}_j) = \exp\left(-\frac{\|\boldsymbol{x}_i - \boldsymbol{x}_j\|^2}{2\sigma^2}\right),$$

where σ^2 is a parameter that adjusts the variance. Because it is difficult to represent graphs as feature vectors without a loss of their structural information, we design a dissimilarity $d(g_i, g_j)$ between g_i and g_j to replace $\|\boldsymbol{x}_i - \boldsymbol{x}_j\|$. A kernel function for graphs is called a graph kernel, denoted by $k(g_i, g_j)$ and defined as

$$k(g_i, g_j) = \exp\left(-\frac{d(g_i, g_j)^2}{2\sigma^2}\right). \tag{1}$$

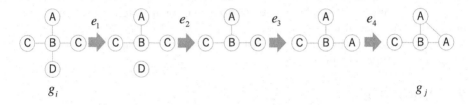

Fig. 2. Sequence of edit operations for transforming g_i into g_j.

3   Mapping Distance Graph Kernels

The definition of $d(g_i, g_j)$ is vital for the performance of the classification model. There are various frameworks for designing graph kernels. Two representative frameworks among them are based on the graph edit distance and graph relabeling. First, we propose a novel graph kernel based on the former framework and then we propose another novel graph kernel based on both frameworks.

3.1   Graph Kernels Based on the Graph Edit Distance

The graph edit distance is one of the most representative metrics for defining $d(g_i, g_j)$, and a number of graph kernels based on this metric have been proposed [21]. The graph edit distance between graphs g_i and g_j is defined as the minimum length of the sequence of edit operations required to transform g_i into g_j, where one edit operation includes the insertion or deletion of a vertex/edge and substitution of a vertex label. Although the edit distance was originally proposed for measuring the dissimilarities between two strings, the metric was extended to graphs because edit operations were introduced for graphs.

Figure 2 shows a particular sequence of edit operations that consists of one deletion of vertex (e_1), one insertion of edge (e_4), one deletion of vertex (e_2), and one substitution of label (e_3). The computation required to obtain the edit distance between g_i and g_j is equivalent to searching for the minimum length of the sequence of edit operations required to transform g_i into g_j. The method based on the $A^\star$ algorithm is a well-known method for computing the exact graph edit distance [16]. However, this method cannot be applied to graphs of large size because the problem of obtaining the exact graph edit distance between two graphs is known to be NP-hard, and consequently, the graph kernels based on the graph edit distance have a drawback in terms of computational efficiency. To address this drawback, we propose a graph kernel based on the mapping distance [23,28], which is the suboptimal graph edit distance between graphs.

Next, we explain the mapping distance between graphs. The distance is one method for approximately measuring the graph edit distance, and this metric is obtained in $O(v^3)$, where $v = \max\{|V(g_i)|, |V(g_j)|\}$. To obtain the mapping distance, we use star structures in the graph [28]. Star structure $s(v)$ for v in graph g is a subtree whose root is v and leaves consist of $N(v)$; that is, $s(v)$ is equivalent to $st(v, 1)$. Given graph g, $|V(g)|$ star structures can be generated from g. The multiset of star structures generated from g is denoted by $S(g) = \{s(v_1), s(v_2), \cdots, s(v_{|V(g)|})\}$. The star edit distance between $s(v_i)$ and $s(v_j)$ is the minimum length of the sequence of edit operations required to transform $s(v_i)$ into $s(v_j)$, and is denoted by $\lambda(s(v_i), s(v_j))$, which is defined as

$$\lambda(s(v_i), s(v_j)) = \lambda_1(v_i, v_j) + \lambda_2(N(v_i), N(v_j)) + \lambda_3(N(v_i), N(v_j)), \qquad (2)$$

where

$$\lambda_1(v_i, v_j) = \delta(\ell(v_i), \ell(v_j)),$$
$$\lambda_2(N(v_i), N(v_j)) = ||N(v_i)| - |N(v_j)||, \text{ and}$$
$$\lambda_3(N(v_i), N(v_j)) = \max\{|N(v_i)|, |N(v_j)|\} - |L(N(v_i)) \cap L(N(v_j))|,$$

where δ is the Kronecker delta that returns one if its arguments are the same, and zero otherwise. Star edit distance $\lambda_1(v_i, v_j)$ returns one if the roots of the star structures have identical labels, and zero otherwise, which is equivalent to a substitution for the labels of roots. Distance $\lambda_2(N(v_i), N(v_j))$ equals the required number of insertions and/or deletions of edges in $s(v_i)$ and $s(v_j)$. Distance $\lambda_3(N(v_i), N(v_j))$ equals the required number of substitutions for the labels of leaves in $s(v_i)$ and $s(v_j)$. From the above, $\lambda(s(v_i), s(v_j))$ represents the star edit distance between $s(v_i)$ and $s(v_j)$.

Given two multisets of star structures $S(g_i)$ and $S(g_j)$, the mapping distance between g_i and g_i is denoted by $md_1(g_i, g_j)$ and defined as

$$md_1(g_i, g_j) = \min_P \sum_{s(u) \in S(g_i)} \lambda(s(u), P(s(u))), \qquad (3)$$

where $P : S(g_i) \rightarrow S(g_j)$ is a bijective function. The computation of $md_1(g_i, g_j)$ is equal to solving the minimum weight matching on the complete bipartite graph $g' = (V_i, V_j, E')$ such that for every two vertices $(v_i, v_j) \in V_i \times V_j$, there is an edge whose weight is the star edit distance $\lambda(s(v_i), s(v_j))$ between $s(v_i)$ and $s(v_j)$. Given a square matrix in which the (i, j)-element represents the star edit distance $\lambda(s(v_i), s(v_j))$, this matching problem is solved using the Hungarian algorithm, which runs in $O(v^3)$ [20], where $v = \max\{|V_i|, |V_j|\}$.

Figure 3 shows an example of mapping $S(g_i)$ to $S(g_j)$ to obtain $md_1(g_i, g_j)$. Given two graphs g_i and g_j, five star structures are generated from g_i and four star structures are generated from g_j. The table at the lower left of the figure represents the star edit distance between every pair of star structures in $S(g_i)$ and $S(g_j)$. Since $|V(g_i)|$ does not equal $|V(g_j)|$, the matrix that represents star edit distances among star structures is not square. To obtain a square matrix, a dummy vertex (denoted by v_5 in g_j) whose label is ϵ is inserted into g_j to equalize the numbers of vertices in g_i and g_j. By applying the Hungarian algorithm, the optimal bipartite graph matching (indicated by solid lines in Fig. 3) is output, and the final answer $md_1(g_i, g_j) = 0 + 2 + 0 + 0 + 5 = 7$ is obtained.

Using the mapping distance, we propose a novel graph kernel called the MDKS.

MDKS: We apply the mapping distance defined in Eq. (3) to the graph kernel defined in Eq. (1). Given two graphs g_i and g_j, the graph kernel in the MDKS is defined as follows:

$$k_{MDKS}(g_i, g_j) = \exp\left(-\frac{md_1(g_i, g_j)^2}{2\sigma^2}\right), \tag{4}$$

where $k_{MDKS}(g_i, g_j)$ is obtained in $O(v^3)$, which is faster than graph kernels based on the exact graph edit distance.

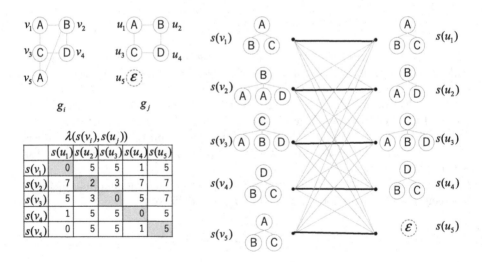

Fig. 3. Minimum weight matching to find the mapping distance between g_i and g_j.

Algorithm 1 shows the pseudocode for computing an MDKS kernel matrix for a set of graphs D. In Lines 2 to 5, the numbers of vertices in g_i and g_j are equalized. For each pair of vertices in $V(g_i) \times V(g_j)$, the star edit distance between star structures $s(v_a)$ and $s(v_b)$ is measured and set as the (a, b)-th element in square matrix T, which is given to the Hungarian algorithm. The

---

Algorithm 1. Mapping_Distance_Kernel1.

Data: a set of graphs D for training and variance σ^2
Result: kernel matrix K

1   for $g_i, g_j \in D$ do
2   while $|V(g_i)| < |V(g_j)|$ do
3        $\lfloor$ $V(g_i) \leftarrow V(g_i) \cup \{dummy\_vertex\}$;
4   while $|V(g_i)| > |V(g_j)|$ do
5        $\lfloor$ $V(g_j) \leftarrow V(g_j) \cup \{dummy\_vertex\}$;
6   for $(v_a, v_b) \in V(g_i) \times V(g_j)$ do
7        $\lambda \leftarrow 0$;
8   if $\ell_i(v_a) \neq \ell_j(v_b)$ then
9          $\lfloor$ $\lambda \leftarrow 1$;
10       $\lambda \leftarrow \lambda + ||N(v_a)| - |N(v_b)||$;
11       $\lambda \leftarrow \lambda + \max\{|N(v_a)|, |N(v_b)|\} - |L(N(v_a)) \cap L(N(v_b))|$;
12       $T_{ab} \leftarrow \lambda$;
13      $md_1 \leftarrow Hungarian(T)$;
14      $K_{ij} \leftarrow \exp\left(-\frac{md_1^2}{2\sigma^2}\right)$;

15 return K;

Hungarian algorithm returns the mapping distance according to the optimal bipartite graph matching in Line 13. In Line 14, Eq. (4) is computed. These procedures are repeated for every pair of graphs in D, and Algorithm 1 finally returns a kernel matrix for D. This algorithm runs in $O(n^2(v^3 + \overline{d}v^2))$, where n, v, and $\overline{d}$ are the number of graphs in D, maximum number of vertices in the graphs, and average degree of the vertices, respectively. Because $\overline{d}$ is bounded by v, the computational complexity becomes $O(n^2v^3)$.

The MDKS has a drawback in terms of graph expressiveness. The height of the subtrees between which we measure the mapping distance is limited. It is desirable to measure the edit distance between high-order subtrees because the edit distance between trees with m vertices is computed in $O(m^3)$ [7]. However, because the paths from the root to leaves in a subtree are not simple in the graph from which the subtree is generated, the number of vertices in the subtrees increases exponentially for h, which makes measuring the edit distance between $s(v_i, h)$ and $s(v_j, h)$ intractable. Another approach [3] is to use subgraphs of g, each of which consists of vertices reachable within h steps from v_i, instead of star structures of G. However, we require exact distances between subgraphs of g_i and subgraphs of g_j, which requires Computation time. In the next subsection, we propose another efficient graph kernel that compares the characteristics of two subtrees, $st(v_i, h)$ and $st(v_j, h)$, for $h > 1$.

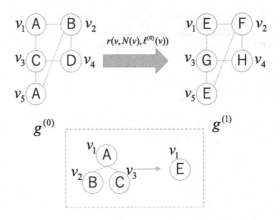

Fig. 4. Example of relabeling ($g^{(0)} \rightarrow g^{(1)}$).

3.2   Graph Kernels Based on Relabeling

Given a graph $g^{(h)} = (V, E, \Sigma, \ell^{(h)})$, all the labels of vertices in $g^{(h)}$ are updated to obtain another graph $g^{(h+1)} = (V, E, \Sigma', \ell^{(h+1)})$. We call the operation a relabel, and it is defined as $\ell^{(h+1)}(v) = r(v, N(v), \ell^{(h)})$. The Weisfeiler–Lehman subtree kernel (WLSK) [27], neighborhood hash kernel (NHK) [11], and Hadamard code kernel (HCK) [13] are representative graph kernels based on this relabeling framework. The vertex label of the WLSK is represented as a string, and a relabel for vertex v is defined as a string concatenation of the labels of $N(v)$. In the NHK, the vertex label is represented as a fixed-length bit string, and relabeling v is defined as logical operations, such as exclusive-or, on the labels of $N(v)$. The label of the HCK is based on the Hadamard code, which is used in spread spectrum-based communication technologies, and a relabel for v is defined as a summation of the labels of $N(v)$.

Figure 4 shows an example of the framework based on graph relabeling. Let $g^{(0)}$ be an original graph whose vertices have labels A, B, C, and D. Each of the labels is relabeled to obtain $g^{(1)}$. Although the specific calculation depends on the method of relabeling, such as the NHK, WLSK, or HCK, it is common that a relabel for v is applied using v, $N(v)$, and $\ell^{(0)}(v)$. At the center of Fig. 4, $\ell^{(0)}(v_1) = A$ is relabeled into E using adjacent vertices v_2, v_3, and its original label A. Therefore, $\ell^{(1)}(v_1) = E$ represents the characteristics of $st(v_1, 1)$. The labels of v_1 and v_5 in $g^{(1)}$ are identical because $st(v_1, 1) = st(v_5, 1)$ in $g^{(0)}$. It is desirable to define labels as identical if and only if both their own labels and the labels of adjacent nodes are also identical. However, achieving this condition is difficult, and it is important to design a relabel method that satisfies this condition as much as possible.

The label of each vertex is relabeled iteratively. Labeling $\ell^{(h)}(v)$, obtained by iteratively relabeling h times, has a distribution of labels that is reachable within h steps from v. Therefore, $\ell^{(h)}(v)$ represents the characteristics of $st(v, h)$. Let $\{g^{(0)}, g^{(1)}, \cdots, g^{(h)}\}$ be a series of graphs obtained by iteratively applying relabeling h times, where $g^{(0)}$ is an original graph contained in D. Kernel $k(g_i, g_j)$ is defined as

$$k(g_i, g_j) = k\left(g_i^{(0)}, g_j^{(0)}\right) + k\left(g_i^{(1)}, g_j^{(1)}\right) + \cdots + k\left(g_i^{(h)}, g_j^{(h)}\right). \qquad (5)$$

The label aggregate kernel (LAK) [13] is another graph kernel based on this framework. Next, we present a specific definition of a relabel in the LAK.

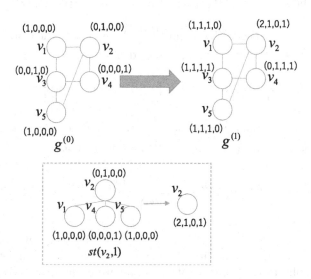

Fig. 5. Example of relabeling in the LAK.

In the LAK, $\ell_L^{(0)}(v)$ is a vector in $|\Sigma|$-dimensional space. If a vertex in a graph has a label σ_i from the set $\Sigma = \{\sigma_1, \sigma_2, \cdots, \sigma_{|\Sigma|}\}$, the i-th element in the vector is one and the other elements are zero. In the LAK, $\ell_L^{(h)}(v)$ is defined as

$$\ell_L^{(h)}(v) = \ell_L^{(h-1)}(v) + \sum_{u \in N(v)} \ell_L^{(h-1)}(u).$$

The i-th element in $\ell_L^{(h)}(v)$ equals the frequency of occurrence of σ_i in $st(v, h)$. Therefore, $\ell_L^{(h)}(v)$ has information on the distribution of labels in $st(v, h)$, which means that $\ell_L^{(h)}(v)$ is more expressive than star structure $s(v)$ used to measure the mapping distance.

We show an example of relabeling in the LAK in Fig. 5, assuming that $|\Sigma| = 4$ and relabeling is applied only once. Consider graph $g^{(0)}$, whose vertices have labels $(1, 0, 0, 0)$, $(0, 1, 0, 0)$, $(0, 0, 1, 0)$, and $(0, 0, 0, 1)$. We next apply relabeling to the graphs to obtain $g^{(1)}$. The label of vertex v in $g^{(1)}$ represents the distribution of labels contained in $st(v, 1)$. For instance, the label of v_2 in $g^{(1)}$ is $\ell_L^{(1)}(v_2) = (2, 1, 0, 1)$, which indicates that there are two vertices labeled

$(1, 0, 0, 0)$, one vertex labeled $(0, 1, 0, 0)$, and one vertex labeled $(0, 0, 0, 1)$. This distribution is equivalent to that of the labels contained in $st(v_2, 1)$ of $g^{(0)}$. In the LAK, the kernel function is defined as

$$k\left(g_i^{(h)}, g_j^{(h)}\right) = \sum_{(v_i, v_j) \in V\left(g_i^{(h)}\right) \times V\left(g_j^{(h)}\right)} \delta\left(\ell_L^{(h)}(v_i), \ell_L^{(h)}(v_j)\right).$$

Using the labels used in the LAK, we propose another novel graph kernel called the MDKV.

MDKV: Given two labels $\ell_L^{(h)}(v_i)$ and $\ell_L^{(h)}(v_j)$, we denote the distance between $\ell_L^{(h)}(v_i)$ and $\ell_L^{(h)}(v_j)$ by $\tau(\ell_L^{(h)}(v_i), \ell_L^{(h)}(v_j))$, defined as

$$\tau\left(\ell_L^{(h)}(v_i), \ell_L^{(h)}(v_j)\right) = \left\|\ell_L^{(h)}(v_i) - \ell_L^{(h)}(v_j)\right\|^2. \tag{6}$$

Given two graphs $g_i^{(h)}$ and $g_j^{(h)}$ relabeled iteratively h times, the distance between $g_i^{(h)}$ and $g_j^{(h)}$ is denoted by $md_2(g_i^{(h)}, g_j^{(h)})$ and defined as

$$md_2\left(g_i^{(h)}, g_j^{(h)}\right) = \min_Q \sum_{u \in V\left(g_i^{(h)}\right)} \tau\left(\ell_L^{(h)}(u), \ell_L^{(h)}(Q(u))\right), \tag{7}$$

where $Q : V(g_i^{(h)}) \to V(g_j^{(h)})$ is a bijective function. The computation of Eq. (7) is also equal to solving the minimum weight matching on a complete bipartite graph, and is obtained by means of the Hungarian algorithm. By combining Eqs. (1), (5), and md_2, kernel $k_{MDKV}(g_i, g_j)$ is defined as follows:

$$k_{MDKV}(g_i, g_j) = \sum_{t=0}^{h} k\left(g_i^{(t)}, g_j^{(t)}\right)$$

$$= \sum_{t=0}^{h} \exp\left(-\frac{md_2\left(g_i^{(t)}, g_j^{(t)}\right)^2}{2\sigma^2}\right).$$

The notable difference between the MDKS and MDKV is that, whereas the inputs for md_1 are multisets of $st(v, 1)$, those for md_2 are multisets of vectors obtained from higher order subtrees $st(v, h)$; that is, the computation of the MDKV between two graphs contains larger subgraphs in the two graphs. If we directly measure the edit distance between $s(v_i, h)$ and $s(v_j, h)$, the number of vertices in $s(v, h)$ exponentially increases when h increases. In this case, the MDKV requires a large amount of Computation time to compute the edit distance. However, using a vector representation for the vertices and their relabeling, our proposed kernel computes a mapping distance between $s(v_i, h)$ and $s(v_j, h)$ efficiently.

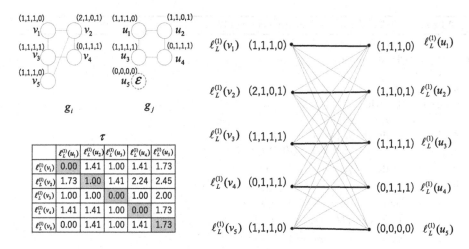

Fig. 6. Computation for obtaining $md_2(g_i^{(1)}, g_j^{(1)})$.

Figure 6 shows an example of the procedure to obtain $md_2(g_i^{(1)}, g_j^{(1)})$, assuming that $|\Sigma| = 4$. Graphs $g_i^{(1)}$ and $g_j^{(1)}$ are obtained by relabeling given graphs $g_i^{(0)}$ and $g_j^{(0)}$ once, respectively. After relabeling, the Euclidean distance between every pair of labels in $g_i^{(2)}$ and $g_j^{(2)}$ is measured. The table between $g_i^{(1)}$ and $g_j^{(1)}$ represents the Euclidean distance between every pair of labels. To equalize the number of vertices in $g_i^{(1)}$ and $g_j^{(1)}$, a dummy vertex whose label is $(0, 0, 0, 0)$ is inserted into $g_j^{(1)}$. The minimum weight matching is solved by means of the Hungarian algorithm, and the final answer of $md_2(g_i^{(1)}, g_j^{(1)}) = 0.00 + 1.00 + 0.00 + 0.00 + 1.73 = 2.73$ is obtained.

Algorithm 2 shows the pseudocode for computing an MDKV kernel matrix for a set of graphs D. In Lines 5 to 8, the numbers of vertices in g_i and g_j are equalized. For each pair of vertices in $V(g_i) \times V(g_j)$, the Euclidean distance between two vectors $\ell_L^{(t)}(v_a)$ and $\ell_L^{(t)}(v_b)$ is measured and set as the (a, b)-th element in T. The Hungarian algorithm returns the mapping distance according to the optimal bipartite graph matching in Line 11. Its output using the Gaussian kernel is added to K_{ij}. In Lines 14 to 16, where $\mathcal{Z}$ is a set of non-negative integers, g is relabeled to obtain $g^{(t+1)}$. These processes in Lines 9 to 15 are repeated $h + 1$ times. This algorithm runs in $O(h(n^2 v^3 + n^2 |\Sigma| v^2 + n\overline{d}|\Sigma| v))$ because the computational complexities of Lines 11, 10, and 15 are $O(v^3)$, $O(|\Sigma| v^2)$, and $O(\overline{d}|\Sigma| v)$, respectively. Because $\overline{d}$ is bounded by v, the computational complexity of Algorithm 2 becomes $O(hn^2(v^3 + |\Sigma| v^2))$.

Algorithm 2. Mapping_Distance_Kernel2.

Data: a set of graphs D for training and variance σ^2
Result: kernel matrix K

1 $K \leftarrow 0$;
2 $D^{(0)} \leftarrow D$;
3 **for** $t \in [0, h]$ **do**
4   **for** $g_i, g_j \in D^{(t)}$ **do**
5   **while** $|V(g_i)| < |V(g_j)|$ **do**
6           $V(g_i) \leftarrow V(g_i) \cup \{dummy\_vertex\}$;
7   **while** $|V(g_i)| > |V(g_j)|$ **do**
8           $V(g_j) \leftarrow V(g_j) \cup \{dummy\_vertex\}$;
9   **for** $(v_a, v_b) \in V(g_i) \times V(g_j)$ **do**
10          $T_{ab} \leftarrow \tau(\boldsymbol{\ell}_L^{(t)}(v_a), \boldsymbol{\ell}_L^{(t)}(v_b))$;
11       $md_2 \leftarrow Hungarian(T)$;
12       $K_{ij} \leftarrow K_{ij} + \exp\left(-\frac{md_2^2}{2\sigma^2}\right)$;
13    $D^{(t+1)} \leftarrow \emptyset$;
14   **for** $g \in D^{(t+1)}$ **do**
15       $g^{(t+1)} \leftarrow (V(g), E(g), \mathcal{Z}^{|\Sigma|}, \ell^{(t+1)})$;
16       $D^{(t+1)} \leftarrow D^{(t+1)} \cup \{g^{(t+1)}\}$;
17 **return** K;

---

4   Evaluation Experiments

In this section, we confirm the Computation time of our proposed graph kernels, the MDKS and MDKV, through numerical experiments. We implemented the proposed graph kernels in Java. All experiments were performed on an Intel Xeon E5-2609 2.50 GHz computer with 32 GB memory running Microsoft Windows 7. To learn from the kernel matrices generated by the above graph kernels, we used the LIBSVM package[1] [4] using 10-fold cross-validation.

Table 1. Parameters of the artificial datasets.

Parameters	Default values
Number of graphs in a dataset	$n = 100$
Average number of vertices in a graph	$\overline{v} = 50$
Average degrees of a graph	$\overline{d} = 2$
Number of distinct labels in a dataset	$|\Sigma| = 10$

---

[1] http://www.csie.ntu.edu.tw/~cjlin/libsvm/.

4.1   Evaluation Using Synthetic Datasets

We examined the computational performance of the proposed graph kernels by means of synthetic graph datasets to confirm that the proposed graph kernels ran in $O(n^2(v^3 + \overline{d}v))$ and $O(h(n^2v^3 + n^2|\Sigma|v^2 + n\overline{d}|\Sigma|v))$, respectively. We generated graphs with a set of four parameters. Their default values are listed in Table 1.

For each dataset, n graphs, each with an average of $\overline{v}$ vertices, were generated. Two vertices in a graph were connected with probability $\frac{\overline{d}}{\overline{v}-1}$, and one label from $|\Sigma|$ was assigned to each vertex in the graph. The Computation times shown in this subsection are the average of 10 trials.

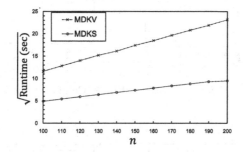

Fig. 7. Computation time for various n.

We first varied only n to generate various datasets in which the other parameters were set to their default values. The number of graphs in each dataset was varied from 10 to 100. Figure 7 shows the Computation time required to generate a kernel matrix for each dataset for the proposed graph kernels[2]. In this experiment, h was set to three. As shown in Fig. 7, the square root of the Computation time for the graph kernels was proportional to the number of graphs in the dataset; that is, Computation time was proportional to the square of the number of graphs in the dataset. This is because the proposed graph kernels were computed for two graphs, and the kernels run for all pairs of graphs in the dataset. We next varied only $\overline{v}$ to generate various datasets with the other parameters set to their default values. Figure 8 shows the Computation time required to generate a kernel matrix in each dataset when the number of vertices in each dataset was varied from 50 to 120. The cube root of the Computation time for the proposed graph kernels was almost proportional to the average number of vertices in the datasets. The parts that required a large amount of Computation time in Algorithms 1 and 2 were those that included the Hungarian algorithm. The Computation time of the algorithm was proportional to the cube of the number of vertices of the bipartite graph that was given as the input. In

---

[2] The figures and tables showing experimental results are the same as ones in the conference version of this paper [14].

the MDKV, $\tau(\ell_L^{(h)}(v_i), \ell_L^{(h)}(v_j))$ represents the dissimilarity between $s(v_i, h)$ and $s(v_j, h)$. The number of vertices in $s(v, h)$ exponentially increased as h increased. If we directly measure the edit distance between $s(v_i, h)$ and $s(v_j, h)$, the MDKV required a large amount of Computation time. However, using a vector representation of the vertices and relabeling the vertices, our proposed MDKV kernel generated the kernel matrix very efficiently.

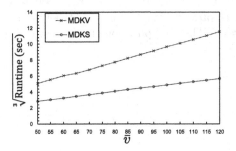

Fig. 8. Computation time for various $\overline{v}$.

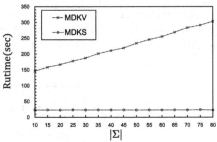

Fig. 9. Computation time for various $|\Sigma|$.

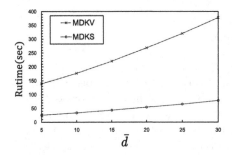

Fig. 10. Computation time for various $\overline{d}$.

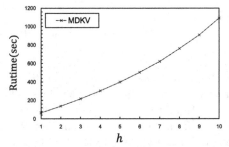

Fig. 11. Computation time for various h.

In Figs. 9 and 10, we varied $|\Sigma|$ and $\overline{d}$, respectively, to generate various datasets. The MDKV required a Computation time that was proportional to $|\Sigma|$ to compute the Euclidean distance $\tau(\ell_L^{(h)}(v_i), \ell_L^{(h)}(v_j))$ and relabel the vertices for the $|\Sigma|$-dimensional vectors. The Computation times for the proposed graph kernels were almost proportional to the average number of degrees of each vertex and number of vertex labels. The MDKS required a Computation time that was proportional to $\overline{d}$ to measure the edit distance of the substitutions for the leaf labels in the star structures. By contrast, the MDKV required a Computation time that was proportional to $\overline{d}$ and $|\Sigma|$ to relabel graphs.

Finally, we varied only h for a dataset generated with all other parameters set to their default values. Figure 11 shows the Computation time required to generate a kernel matrix in each dataset when h was varied from 0 to 15. The Computation time was proportional to h.

4.2   Classification Accuracy

We compared the classification accuracies of the proposed graph kernels with those of conventional graph kernels based on the relabeling framework, WLSK, NHK, and HCK on three real-world datasets: MUTAG [6], PTC [10], and ENZYMES [25]. Because the HCK theoretically returned the same values as the LAK, the classification accuracies of the HCK were equivalent to those of the LAK. The first dataset MUTAG consists of 188 chemical compounds, and their classes are binary values that represent whether each compound is mutagenic. The second dataset PTC consists of 344 chemical compounds, and their classes are binary values that represent whether each compound is toxic. Generally, a chemical compound is represented as a graph with labeled edges, which is not a graph that is considered in this paper. We considered the graphs with edge labels using the following two approaches: (1) we ignored the edge labels; or (2) an edge labeled ℓ that was adjacent to vertices u and v in a graph was converted into a vertex labeled ℓ that was adjacent to u and v, as explained in [11]. After converting the edges in graphs, labels were assigned to only the vertices. The third dataset, ENZYMES consists 600 proteins, and their classes represent Enzyme Commission numbers from one to six. Table 2 shows a summary of each dataset.

Before classifying a dataset that does not contain graphs but consists of points in a p-dimensional feature space, we typically normalize the dataset using the mean μ_q and standard deviation σ_q in the q-th feature ($1 \leq q \leq p$). By normalizing the dataset, we often obtain an accurate model for classifying the dataset. Similarly, we applied this procedure in the MDKV using the mean $\mu_q^{(t)}$ and standard deviation $\sigma_q^{(t)}$ for the $|\Sigma|$-dimensional vectors to represent the vertex labels for each t ($1 \leq t \leq h$) and q ($1 \leq q \leq |\Sigma|$). Using this procedure, we avoided the exponential increase in the elements in the vectors that represented vertex labels when h increased. We call the MDKV method that uses this procedure St-MDKV.

Table 3 shows the classification accuracies of the proposed and conventional graph kernels. We examined the highest accuracy for each kernel and each dataset, varying σ of the Gaussian kernel and h. We varied σ from σ_{min} to σ_{max} in intervals of 10 (see Table 4) and h from zero to 15 in intervals of one. As shown in Table 3, the classification accuracies of the proposed graph kernels outperformed those of conventional graph kernels. The values of h for the MDKV were relatively low, which indicates that the elements in the vectors representing vertex labels exponentially increased and distance $\tau(\ell_L^{(h)}(v_i), \ell_L^{(h)}(v_j))$ became inadequate when h increased. However, the values of h for St-MDKV were high. By normalizing the vectors representing vertex labels, we adequately measured the (dis)similarity between $s(v_i, h)$ and $s(v_j, h)$, which resulted in high classification accuracy for various datasets.

Table 2. Description of the evaluation datasets.

	MUTAG		PTC		ENZYMES
	Edge labels	No edge labels	Edge labels	No edge labels	
Number of graphs n	188		344		600
Number of classes (class distribution)	2 (125,63)		2 (152,192)		6 (100 per class)
Max. number of vertices	84	40	325	109	126
Avg. number of vertices	53.9	26.0	77.5	25.6	32.6
Number of labels	12	8	67	19	3
Average degree	2.1	2.1	2.7	4.0	3.9

5 Discussion on Related Work

5.1 Graph Edit Distance Based on the Mapping Distance

The proposed graph kernels in this paper are related to methods for computing the graph edit distance. This subsection surveys a framework for computing the approximate graph edit distance between two graphs using the linear sum assignment problem (LSAP).

Table 3. Classification accuracies.

	MUTAG		PTC		ENZYMES
	Edge labels	No edge labels	Edge labels	No edge labels	
MDKS	92.6%	91.0%	64.2%	63.1%	61.2%
MDKV	**94.1%** $(h=3)$	**93.6%** $(h=2)$	64.0% $(h=0)$	**66.9%** $(h=3)$	**65.3%** $(h=2)$
St-MDKV	91.5% $(h=7,8)$	90.4% $(h=3)$	**64.9%** $(h=1,8)$	64.0% $(h=1)$	63.0% $(h=4)$
NHK	92.6% $(h=3,4)$	90.4% $(h=2)$	60.8% $(h=3,5)$	55.8% $(h=1,2,\cdots,15)$	45.0% $(h=8)$
WLSK	92.0% $(h=3)$	90.4% $(h=1)$	62.8% $(h=15)$	64.2% $(h=10)$	58.5% $(h=1)$
HCK	92.0% $(h=3)$	91.0% $(h=1)$	63.1% $(h=15)$	65.4% $(h=12)$	57.2% $(h=4)$

The problem of computing the exact graph edit distance between graphs $g_1 = (V_1, E_1, \ell_1)$ and $g_2 = (V_2, E_2, \ell_2)$ is formalized as follows [24]: The sets of vertices V_1 and V_2 are extended to

Table 4. Description of the evaluation settings.

	MUTAG	PTC	ENZYMES
σ_{min}	10	10	10^2
σ_{max}	10^4	10^5	10^4

$$V_1^+ = V_1 \cup \overbrace{\{\,\varepsilon_1, \varepsilon_2, \cdots, \varepsilon_b\,\}}^{b \text{ empty vertices}} \text{ and}$$

$$V_2^+ = V_2 \cup \overbrace{\{\,\varepsilon_1, \varepsilon_2', \cdots, \varepsilon_a'\,\}}^{a \text{ empty vertices}},$$

respectively, where $|V_1| = a$ and $|V_2| = b$. The graph edit distance computation is eventually performed on graphs $g_1 = (V_1^+, E_1, \ell_1)$ and $g_2 = (V_2^+, E_2, \ell_2)$. Additionally, a cost matrix for editing g_1 to g_2 is defined as

$$C = \begin{array}{c} \\ 1 \\ 2 \\ \vdots \\ a \\ 1 \\ 2 \\ \vdots \\ b \end{array} \begin{array}{c} \begin{array}{cccc} 1 & 2 & \cdots & b \end{array} \begin{array}{cccc} 1 & 2 & \cdots & a \end{array} \\ \left(\begin{array}{cccc|cccc} c_{11} & c_{12} & \cdots & c_{1b} & c_{1\varepsilon} & \infty & \cdots & \infty \\ c_{21} & c_{22} & \cdots & c_{2b} & \infty & c_{2\varepsilon} & \ddots & \vdots \\ \vdots & \vdots & \ddots & \vdots & \vdots & \ddots & c_{1\varepsilon} & \infty \\ c_{a1} & c_{a2} & \cdots & c_{ab} & \infty & \cdots & \infty & c_{a\varepsilon} \\ \hline c_{\varepsilon 1} & \infty & \cdots & \infty & 0 & 0 & \cdots & 0 \\ \infty & c_{\varepsilon 2} & \ddots & \vdots & 0 & 0 & \cdots & 0 \\ \vdots & \ddots & \ddots & \infty & \vdots & \vdots & \ddots & \vdots \\ \infty & \cdots & \infty & c_{\varepsilon b} & 0 & 0 & \cdots & 0 \end{array} \right) \end{array} \quad (8)$$

where c_{ij}, $c_{i\varepsilon}$, and $c_{\varepsilon i}$ denote the costs of replacing a label of vertex $v_i \in V_1$ with a label of vertex $v_j \in V_2$, deleting a vertex v_i from g_1, and inserting a vertex v_i into g_1 to edit g_1 to g_2, respectively. Additionally, we denote the cost of editing an edge (v_i, v_j) in g_1 into an edge $(v_{i'}, v_{j'})$ in g_2 by $c((v_i, v_j) \to (v_{i'}, v_{j'}))$. Because we assume that only vertices in graphs have labels in this paper, $c((v_i, v_j) \to (v_{i'}, v_{j'}))$ is the cost of inserting or deleting an edge. When g_1 is edited to g_2, the mapping between the two extended sets of vertices is denoted by a bijective function $\varphi : V_1^+ \to V_2^+$. Then, the total cost of editing g_1 to g_2 via φ is

$$dist(g_1, g_2, \varphi) = \sum_{i=1}^{a+b} c_{i\varphi(i)} + \sum_{i=1}^{a+b} \sum_{j=i+1}^{a+b} c\left((v_i, v_j) \to (v_{\varphi(i)}, v_{\varphi(j)})\right). \quad (9)$$

Therefore, the exact edit distance between g_1 and g_2 is

$$dist(g_1, g_2) = \min_{\varphi \in \Phi} dist(g_1, g_2, \varphi), \quad (10)$$

where Φ is a set of all possible permutations of integers $1, 2, \cdots, a + b$. Equation (10) is a type of quadratic assignment problem that is known to be NP-complete [15], although it is the LSAP if Eq. (9) does not contain its second term.

Risen et al. proposed an algorithmic and efficient framework that enables us to obtain the approximate graph edit distance by omitting the second term of Eq. (9). The framework first solves

$$\hat{\varphi} = \arg\min_{\varphi \in \varPhi} \sum_{i=1}^{a+b} c_{i\varphi(i)} \qquad (11)$$

and then obtains the approximate edit distance between g_1 and g_2 by substituting $\hat{\varphi}$ into Eq. (9). Equation (11) implies that each vertex v_i in g_1 should be mapped to a vertex in g_2 that has the same label as the label of v_i as much as possible. Solving Eq. (11) is equivalent to the minimum matching problem of a bipartite graph whose vertices are V_1^+ and V_2^+, and whose edge weights are c_{ij} of Eq. (8). Therefore, this problem is tractable in $O((a+b)^3)$. However, because the structural information of the two graphs is ignored and only vertex labels are taken into account in the optimization problem shown in Eq. (9), we do not always obtain an adequate mapping from V_1^+ to V_2^+. To overcome this difficulty, the elements of the cost matrix of Eq. (8) are redefined to take account of the structural information as

$$c_{ij}^* = c_{ij} + c(local(v_i) \rightarrow local(v_j)),$$

where $local(v_i)$ is local structure around a vertex v_i, and $c(local(v_i) \rightarrow local(v_j))$ is the cost of editing $local(v_i)$ to $local(v_j)$. Recently, various methods in this framework have been proposed to represent local structures around vertices to obtain a more accurate mapping between sets of vertices than that of Eq. (11).

In [9], $local(v_i)$ is a set of walks with k steps from vertex v_i. Each walk is a label sequence of length $2k + 1$. Two independent walks on two graphs are efficiently obtained from a walk on a product graph for the two graphs. The product graph $g_\times = (V_\times, E_\times)$ of the two graphs is defined as

$$V_\times = \{(v_i, v_j) \in V_1 \times V_2 \mid \ell_1(v_i) = \ell_2(v_j)\} \text{ and}$$
$$E_\times = \{((v_i, v_j), (v_{i'}, v_{j'})) \in V_\times \times V_\times \mid$$
$$(v_i, v_{i'}) \in E_1 \wedge (v_j, v_{j'}) \in E_2 \wedge \ell_1((v_i, v_{i'})) = \ell_2((v_j, v_{j'}))\}.$$

When W is an adjacency matrix of the product graph whose elements are binary, the number of walks that generate identical label sequences is obtained from an element of W^k. The edit distance based on a walk leads to the problem of "tottering," that is, iteratively visiting the same cycle of vertices; a walk can generate artificially low edit costs (high similarity values) [2].

In [23,24], $local(v_i)$ is defined by vertex v_i and its adjacent edges as

$$c_{ij}^* = c_{ij} + \min_{\varphi \in \varPhi} \sum_{k=1, k\neq i,j}^{a+b} c((v_i, v_k) \rightarrow (v_j, v_{\varphi(k)})). \qquad (12)$$

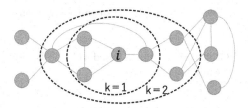

Fig. 12. Limited-size subgraphs induced by vertices within k steps from vertex v_i.

This equation implies that each vertex v_i in g_1 should be mapped to a vertex in g_2 that has the same label as the label of v_i and the same number of edges as that for v_i as much as possible. Because c_{ij}^* of Eq. (12) is computable in $O(a+b)$, the matrix C^* whose elements are c_{ij}^* is computable in $O((a+b)^3)$. Therefore, the overall computation to solve Eqs. (12), (11), and (9) is $O((a+b)^3)$. As explained in Sect. 3.1, [28] defines $local(v_i)$ as a vertex v_i, its adjacent edges, and vertices. In this method, c_{ij}^* is defined as

$$c_{ij}^* = \lambda((s(i), s(j))) = c_{ij} + \lambda_2((s(i), s(j))) + \lambda_3((s(i), s(j))), \qquad (13)$$

where λ, λ_2, and λ_3 are defined in Eq. (2). Compared with Eqs. (12), (13) is efficient because it does not contain the optimization problem.

In [3], $local(v_i)$ is a limited-size subgraph induced by vertices reachable within k steps from vertex v_i, as shown in Fig. 12. The subgraph $local(v_i)$ is denoted by $\mathcal{N}_k^i$, and c_{ij}^* is the exact edit distance between $\mathcal{N}_k^i$ and $\mathcal{N}_k^j$, that is, $c_{ij}^* = dist(\mathcal{N}_k^i, \mathcal{N}_k^j)$. Because the problem of a graph edit distance is NP-complete, although applying the problem to large graphs is intractable, $dist(\mathcal{N}_k^i, \mathcal{N}_k^j)$ is tractable for small k. Additionally, because the structural information that limited-size subgraphs have is more than that of walks and stars, the graph edit distance based on limited-size subgraphs provides an accurate approximate graph edit distance.

Table 5 summarizes citations in which methods for computing the approximate graph edit distance are proposed, as provided in this subsection. The next subsection discusses some graph kernels.

5.2 Support Vector Machines with Graph Kernels

Given a set of examples $D = \{(\boldsymbol{x}_i, y_i)\}_{i=1}^n$ such that $\boldsymbol{x}_i \in \mathcal{R}^p$ and $y_i \in \{-1, +1\}$, the SVM determines a hyperplane $\boldsymbol{w}^T\boldsymbol{x} + b = 0$ in $\mathcal{R}^p$ that satisfies $y_i(\boldsymbol{w}^T\boldsymbol{x}_i + b) \geq 1$. Some examples in D that are the nearest to the hyperplane are called support vectors, and the distance between the support vectors and hyperplane is $\frac{1}{||\boldsymbol{w}||}$. Because the hyperplane that classifies the positive and negative examples in D accurately is such that the margin $\frac{1}{||\boldsymbol{w}||}$ is maximum under $y_i(\boldsymbol{w}^T\boldsymbol{x}_i + b) \geq 1$, this problem is formalized as follows:

Table 5. Methods for measuring the graph edit distance based on the mapping distance and graph kernels.

	Graph edit distance based on mapping distance	Graph kernel
Path, walk	Gaüzère [9]	Kashima [12], Borgwardt [2], Gärtner [8]
Star	Riesen [23], Zeng [28]	–
Tree	–	Mahé [22], Bach [1]
Limited-size graph	Carletti [3]	Costa [5], Horváth [18], Shervashidze [26]
Relabel	–	Hido [11], Shervashidze [27], Kataoka [11]

$$\min_{\boldsymbol{w},b} \ ||\boldsymbol{w}||^2$$

$$s.t. \ y_i(\boldsymbol{w}^T \boldsymbol{x}_i + b) \geq 1.$$

By applying the method of Lagrange multipliers to the above optimization problem, we obtain the following dual problem:

$$\max_{\alpha_1,\cdots,\alpha_n} \ -\frac{1}{2}\sum_{i=1}^{n}\sum_{j=1}^{n}\alpha_i\alpha_j y_i y_j k(\boldsymbol{x}_i,\boldsymbol{x}_j) + \sum_{i=1}^{n}\alpha_i$$

$$s.t. \ \alpha_i \geq 0, \ \sum_{i=1}^{n}\alpha_i y_i = 0,$$

where α_i are Lagrange multipliers and $k(\boldsymbol{x}_i,\boldsymbol{x}_j) = \boldsymbol{x}_i^T \boldsymbol{x}_j$. Using the Lagrange multipliers, the obtained hyperplane is denoted by

$$f(\boldsymbol{x}) = sign\left(\sum_{i=1}^{n}\alpha_i y_i k(\boldsymbol{x}_i,\boldsymbol{x}) + b\right).$$

In this paper, we consider a problem for classifying labeled graphs accurately. A straightforward method for converting a set of labeled graphs to a set of feature vectors in $\mathcal{R}^p$ to apply the conventional SVM is as follows: First, we enumerate frequent subgraph patterns F from the set of labeled graphs G in the training dataset, where the frequent subgraph patterns F are defined as graphs f appearing in G more frequently than threshold μ; that is, $sup(f) = |\{g \mid g \in G, f \subseteq g\}|$ and $F = \{f \mid sup(f) \geq \mu\}$ [19]. Then, we derive a feature vector $\boldsymbol{x} = (x_1, x_2, \cdots, x_{|F|})^T \in \mathcal{R}^{|F|}$ for a graph g in G, where x_j is one if $f_j \in F$ and $f_j \subseteq g$, and zero otherwise. Based on this conversion, because all the graphs are represented as feature vectors, we can apply the conventional SVM with kernels, such as polynomial, Gaussian, and sigmoid kernels, to the feature vectors. However, to enumerate all frequent subgraph patterns from a set of graphs requires much Computation time, and the subgraph isomorphism problem $f_j \subseteq g$ in which a pattern f_j is included in graph g is known to be NP-complete. Additionally, because we cannot generate graph g from its feature vector x, there is loss of structural information by converting g to x, which may cause a decrease in classification accuracy on constructing an SVM.

Since the introduction of convolution kernels in [17], the decomposition approach has been the guiding principle in kernel design for structured objects. According to such an approach, a similarity function between discrete data structures can be obtained by decomposing each object into parts and devising a valid local kernel between the subparts [5]. Thus, in most graph kernels $k(g_i, g_j)$, graphs g_i and g_j are decomposed into subparts $local(v)$ and $local(u)$, respectively, and then local kernels for the subparts are accumulated, such as $k(g_i, g_j) = \sum_{v \in g_i} \sum_{u \in g_j} k(local(v), local(u)) = \sum_{v \in g_i} \sum_{u \in g_j} k(v, u)$.

According to [27], the first class of graph kernels is based on random walks. In [12], $k(v, u)$ is defined by sets of random walks, with k steps from vertices v and u. A walk from v is denoted by a sequence of vertex labels and edge labels $w_v = \ell_{v_0} \ell_{v_0,v_1} \ell_{v_1} \ell_{v_1,v_2} \cdots \ell_{v_{k-1}} \ell_{v_{k-1},v_k} \ell_{v_k}$, where ℓ_v is a label of v and $\ell_{v,u}$ is a label of (v, u). For this walk, a random walk kernel is defined as

$$k(v, u) = p(w_v) p(w_u) \delta(\ell_{v_0}, \ell_{u_0}) \prod_{h=1}^{k} \delta(\ell_{v_{h-1},v_h}, \ell_{u_{h-1},u_h}) \delta(\ell_{v_h}, \ell_{u_h}), \qquad (14)$$

where $p(w_v)$ is the probability that a walk w_v occurs. To avoid deriving walks w_v and w_u such that $\delta(\ell_{v_0}, \ell_{u_0}) \prod_{h=1}^{k} \delta(\ell_{v_{h-1},v_h}, \ell_{u_{h-1},u_h}) \delta(\ell_{v_h}, \ell_{u_h}) = 0$, which is the underlined part of Eq. (14), and raise its computational efficiency, a product graph of g_i and g_j is used [2].

The second class of graph kernels is based on trees. For example, in [22], the size-based balanced tree-pattern kernel is defined as

$$k(v, u, h) = \sum_{t \in \mathcal{B}_h} \lambda^{|t|-h} \psi(g_i, v, t) \psi(g_j, u, t), \qquad (15)$$

where $\mathcal{B}_h$ is a set of balanced trees of height h, $|t|$ is the size of t, λ is a parameter, and $\psi(g, v, t)$ is a tree count function that returns the number of times that a tree-pattern t occurs in a graph g, where the root of t is mapped to v in g. Using dynamic programing, Eq. (15) is rewritten as

$$k(v, u, h) = \lambda \delta(\ell_v, \ell_u) \sum_{R \in \mathcal{M}(v,u)} \prod_{(v',u') \in R} k(v', u', h - 1),$$

where $k(v, u, 1) = \lambda \delta(\ell_v, \ell_u)$ and $\mathcal{M}$ is defined as

$$\mathcal{M}(v, u) = \{R \subseteq N(v) \times N(u) \mid$$
$$\forall (a, b) \in R, \ell(a) = \ell(b) \wedge \ell(v, a) = \ell(u, b)\}.$$

Each $R \in \mathcal{M}(v, u)$ consists of one or several pairs of neighbors of v and u that are identically labeled and connected to v and u by edges of the same label.

The third class of graph kernels is based on limited-size subgraphs, including kernels based on so-called graphlets [26], which represent graphs as counts of all types of subgraphs of size $k \in \{3, 4, 5\}$. For example, in [5], the neighborhood subgraph pairwise distance kernel is defined as follows using $\mathcal{N}_k^v$, which is a limited-size subgraph induced by vertices reachable within k steps from vertex v:

$$k(v, u) = \sum_{d>0} \sum_{k} \sum_{\substack{v' \in V(g_i) \\ dist(v,v')=d}} \sum_{\substack{u' \in V(g_j) \\ dist(u,u')=d}} \delta(\mathcal{N}_k^v, \mathcal{N}_k^{v'}) \delta(\mathcal{N}_k^u, \mathcal{N}_k^{u'}),$$

where $dist(v, v')$ is the shortest step between vertices v and v' in a graph. This kernel counts all pairs of neighborhood graphs of radius k whose roots are at distance d. It requires graph isomorphism matching for small graphs $\mathcal{N}_k^v$ and $\mathcal{N}_k^{v'}$ in $\delta(\mathcal{N}_k^v, \mathcal{N}_k^{v'})$.

The fourth class of graph kernels are based on relabeling. In these kernels, $\mathcal{N}_k^v$ for a vertex v is represented as a label but not a small graph, so we can quickly check whether two labels $\mathcal{N}_k^v$ and $\mathcal{N}_k^u$ are equivalent. Given graph $g^{(h)} = (V, E, \Sigma, \ell^{(h)})$, a procedure that converts $g^{(h)}$ to another graph $g^{(h+1)} = (V, E, \Sigma', \ell^{(h+1)})$ is called a relabel. The label of a vertex v in $g^{(h+1)}$ is defined using the labels of v and $N(v)$ in $g^{(h)}$, and is denoted by $\ell^{(h+1)}(v) = r(v, N(v), \ell^{(h)})$. Let $\{g^{(0)}, g^{(1)}, \cdots, g^{(h)}\}$ be a series of graphs obtained by iteratively applying a relabel h times, where $g^{(0)}$ is one of graphs in the dataset. Given two graphs g_i and g_j, a graph kernel is defined as

$$k(v, u) = \sum_{t=1}^{h} \delta\left(\ell_v^{(t)}, \ell_u^{(t)}\right), \tag{16}$$

where $\ell_v^{(t)}$ is a label for v in $g^{(t)}$. Recently, various graph kernels, such as the NHK [11], WLSK [27], LAK [13], HCK [13], and shortened Hadamard code kernel [13], have been proposed to apply to the graph classification problem.

In this paper, we proposed two graph kernels called MDKS and MDKV. MDKS is based on the graph edit distance using star structures. Because the kernel uses only small local structures that consist of adjacent vertices of each vertex in graphs, it is not substantially superior to conventional graph kernels. MDKV is based on the framework of the graph edit distance using LSAP and the framework of relabeling in conventional graph kernels. The kernel uses local structures that consist of vertices that are reachable within h steps from each vertex in graphs and do not require isomorphism matching similar to, for example, the NHK, WLSK, and HCK. Additionally, similarities between local structures are represented as continuous values by Eq. (6) but not binaries by the underlined part of Eq. (16), which enables us to measure similarities between graphs adequately. This is why our proposed graph kernel MDKV is substantially superior to conventional graph kernels.

6   Conclusion

In this paper, we proposed two novel graph kernels, mapping distance kernel with stars (MDKS) and mapping distance kernel with vectors (MDKV), to classify labeled graphs more accurately than existing methods. The MDKS is based on the graph edit distance using star structures, and the MDKV is based on the graph edit distance using the linear sum assignment problem and graph relabeling. To verify the computational efficiency of the proposed graph kernels, we performed experiments on an artificially generated dataset. Additionally, we compared the classification accuracy of the proposed graph kernels with conventional graph kernels using real-world datasets. Because MDKS uses only small local structures that consist of adjacent vertices of each vertex in graphs, it is not substantially superior to conventional graph kernels. However, the MDKV uses local structures that consist of vertices that are reachable within a small number of steps from each vertex in graphs and, unlike existing methods, do not require isomorphism matching. Thus, the MDKV is substantially superior to conventional graph kernels.

References

1. Bach, F.R.: Graph kernels between point clouds. In: Proceedings of the International Conference on Machine Learning (ICML), pp. 25–32 (2008)
2. Borgwardt, K.M., Kriegel, H.-P.: Shortest-path kernels on graphs. In: Proceedings of IEEE International Conference on Data Mining (ICDM), pp. 74–81 (2005)
3. Carletti, V., Gaüzère, B., Brun, L., Vento, M.: Approximate graph edit distance computation combining bipartite matching and exact neighborhood substructure distance. In: Liu, C.-L., Luo, B., Kropatsch, W.G., Cheng, J. (eds.) GbRPR 2015. LNCS, vol. 9069, pp. 188–197. Springer, Cham (2015). https://doi.org/10.1007/978-3-319-18224-7_19
4. Chang, C.-C., Lin, C.-J.: LIBSVM: A Library for Support Vector Machines (2001). http://www.csie.ntu.edu.tw/cjlin/libsvm
5. Costa, F., De Grave, K.: Fast neighborhood subgraph pairwise distance kernel. In: Proceedings of International Conference on Machine Learning (ICML), pp. 255–262 (2010)
6. Debnath, A.K., Lopez de Compadre, R.L., Debnath, G., Shusterman, A.J., Hansch, C.: Structure-activity relationship of mutagenic aromatic and heteroaromatic nitro compounds. Correlation with molecular orbital energies and hydrophobicity. J. Med. Chem. **34**, 786–797 (1991)
7. Demaine, E.D., Mozes, S., Rossman, B., Weimann, O.: An optimal decomposition algorithm for tree edit distance. ACM Trans. Algorithm **6**(1), 2:1–2:19 (2009)
8. Gärtner, T., Flach, P., Wrobel, S.: On graph kernels: hardness results and efficient alternatives. In: Schölkopf, B., Warmuth, M.K. (eds.) COLT-Kernel 2003. LNCS (LNAI), vol. 2777, pp. 129–143. Springer, Heidelberg (2003). https://doi.org/10.1007/978-3-540-45167-9_11
9. Gaüzère, B., Bougleux, S., Riesen, K., Brun, L.: Approximate graph edit distance guided by bipartite matching of bags of walks. In: Fränti, P., Brown, G., Loog, M., Escolano, F., Pelillo, M. (eds.) S+SSPR 2014. LNCS, vol. 8621, pp. 73–82. Springer, Heidelberg (2014). https://doi.org/10.1007/978-3-662-44415-3_8

10. Helma, C., Kramer, S.: A survey of the predictive toxicology challenge. Bioinformatics **19**(10), 1179–1182 (2003)
11. Hido, S., Kashima, H.: A linear-time graph kernel. In: Proceedings of the International Conference on Data Mining (ICDM), pp. 179–188 (2009)
12. Kashima, H., Tsuda, K., Inokuchi, A.: Marginalized kernels between labeled graphs. In: Proceedings of the International Conference on Machine Learning (ICML), pp. 321–328 (2003)
13. Kataoka, T., Inokuchi, A.: Hadamard code graph kernels for classifying graphs. In: Proceedings of the International Conference on Pattern Recognition Applications and Methods (ICPRAM), pp. 24–32 (2016)
14. Kataoka, T., Shiotsuki, E., Inokuchi, A.: Mapping distance graph kernels using bipartite matching. In: Proceedings of the International Conference on Pattern Recognition Applications and Methods (ICPRAM), pp. 61–70 (2017)
15. Koopmans, T.C., Beckmann, M.: Assignment problems and the location of economic activities. Econometrica **25**(1), 53–76 (1957)
16. Hart, P.E., Nilsson, N.J., Raphael, B.: A formal basis for the heuristic determination of minimum cost paths. IEEE Trans. Syst. Sci. Cybern. **4**(2), 100–107 (1968)
17. Haussler, D.: Convolution kernels on discrete structures. Technical report, UCSC-CRL-99-10, University of California at Santa Cruz (1999)
18. Horváth, T., Gärtner, T., Wrobel, S.: Cyclic pattern kernels for predictive graph mining. In: Proceedings of the ACM SIGKDD Conference on Knowledge Discovery and Data Mining, (KDD), pp. 158–167 (2004)
19. Inokuchi, A., Washio, T., Motoda, H.: An apriori-based algorithm for mining frequent substructures from graph data. In: Zighed, D.A., Komorowski, J., Żytkow, J. (eds.) PKDD 2000. LNCS (LNAI), vol. 1910, pp. 13–23. Springer, Heidelberg (2000). https://doi.org/10.1007/3-540-45372-5_2
20. Kuhn, H.W.: The Hungarian method for the assignment problem. Nav. Res. Logist. **2**, 83–97 (1955)
21. Neuhaus, M., Bunke, H.: Bridging the Gap Between Graph Edit Distance and Kernel Machines. World Scientific, Singapore (2007)
22. Mahé, P., Vert, J.-P.: Graph kernels based on tree patterns for molecules. Mach. Learn. **75**(1), 3–35 (2009)
23. Riesen, K., Bunke, H.: Approximate graph edit distance computation by means of bipartite graph matching. Image Vis. Comput. **27**(7), 950–959 (2009)
24. Riesen, K.: Structural Pattern Recognition with Graph Edit Distance. ACVPR. Springer, Cham (2015). https://doi.org/10.1007/978-3-319-27252-8
25. Schomburg, I., Chang, A., Ebeling, C., Gremse, M., Heldt, C., Huhn, G., Schomburg, D.: BRENDA, the enzyme database: updates and major new developments. Nucleic Acids Res. **32D**, 431–433 (2004)
26. Shervashidze, N., Vishwanathan, S.V.N., Petri, T., Mehlhorn, K., Borgwardt, K.M.: Efficient graphlet kernels for large graph comparison. In: Proceedings of the Twelfth International Conference on Artificial Intelligence and Statistics (AISTATS), pp. 488–495 (2009)
27. Shervashidze, N., Schweitzer, P., van Leeuwen, E.J., Mehlhorn, K., Borgwardt, K.M.: Weisfeiler-Lehman graph kernels. J. Mach. Learn. Res. (JMLR) **12**, 2539–2561 (2011)
28. Zeng, Z., Tung, A.K.H., Wang, J., Feng, J., Zhou, L.: Comparing stars: on approximating graph edit distance. Proc. Int. Conf. Very Large Database (VLDB) **2**(1), 25–36 (2009)

Domain Adaptation Transfer Learning by Kernel Representation Adaptation

Xiaoyi Chen$^{(\boxtimes)}$ and Régis Lengellé

Université de Technologie de Troyes, ROSAS Department,
Charles Delaunay Institute, UMR CNRS 6281, LM2S Research Team,
12 rue Marie Curie, CS 42060, 10004 Troyes Cedex, France
`xiaoyi.chen@utt.fr`

Abstract. Domain adaptation, where no labeled target data is available, is a challenging task. To solve this problem, we first propose a new SVM based approach with a supplementary Maximum Mean Discrepancy (MMD)-like constraint. With this heuristic, source and target data are projected onto a common subspace of a Reproducing Kernel Hilbert Space (RKHS) where both data distributions are expected to become similar. Therefore, a classifier trained on source data might perform well on target data, if the conditional probabilities of labels are similar for source and target data, which is the main assumption of this paper. We demonstrate that adding this constraint does not change the quadratic nature of the optimization problem, so we can use common quadratic optimization tools. Secondly, using the same idea that rendering source and target data similar might ensure efficient transfer learning, and with the same assumption, a Kernel Principal Component Analysis (KPCA) based transfer learning method is proposed. Different from the first heuristic, this second method ensures other higher order moments to be aligned in the RKHS, which leads to better performances. Here again, we select MMD as the similarity measure. Then, a linear transformation is also applied to further improve the alignment between source and target data. We finally compare both methods with other transfer learning methods from the literature to show their efficiency on synthetic and real datasets.

1 Introduction

With the trend of Artificial Intelligence, there is more need to transfer knowledge from what we have trained to another (sometimes more difficult) task (task to be trained), especially when the latter is different yet related to the trained task. This is the main objective of *Transfer Learning*: taking full advantage of previous knowledge to learn a good classifier or regressor in a new different but related domain, where learning may be much more difficult using only the new domain. We designate previous knowledge as source, which is supposed available, while the new domain is designated as target. Depending on the availability of labels of target and source domains, transfer learning can be categorized as multi-task

© Springer International Publishing AG, part of Springer Nature 2018
M. De Marsico et al. (Eds.): ICPRAM 2017, LNCS 10857, pp. 45–61, 2018.
https://doi.org/10.1007/978-3-319-93647-5_3

learning, self-taught learning, transductive transfer learning and unsupervised transfer learning (according to [19]). In this paper, we focus on transductive transfer learning, where there is labeled source data and no labeled target data. In this sub-branch, we consider that source and target data distributions are different in their feature space but share the same label space. Marginal, conditional distributions of observations or priors can be different. This problem belongs to domain adaptation problems.

There is a variety of transfer learning methods. In this paper, we suppose that the conditional probabilities of labels of source and transformed target data, given an observation, are similar in a Reproducing Kernel Hilbert Space (RKHS) subspace, that is:

$$\exists g(.) : \mathbb{R}^m \to \mathbb{R}^n \mid p_s(y|x, x \in \mathcal{S}) = p_s(y|g(x), x \in \mathcal{T})$$

where $g(.)$ is a smooth transformation function, y represents the label, x is an observation either taken from the source domain ($\mathcal{S}$) or from the target domain ($\mathcal{T}$).

Within this assumption, we first propose the use of a *Support Vector Machine (SVM)* subject to a zero valued *Maximum Mean Discrepancy (MMD)*-like constraint, then we consider extending the principal idea to solve the domain adaptation problem using *Kernel Principal Component Analysis (KPCA)*.

For the first *MMD-like constrained SVM* approach, the reason of the choice of a zero-valued MMD-like constraint is that MMD is a non-parametric measure of the distance between 2 distributions [4] and it can be easily kernelized [8]. Therefore, combining MMD and SVM appears promising. SVM is a famous method used in binary classification. It is well known for its generalization ability and the simplicity in dealing with non-linearly separable data set. Our method keeps these advantages while performing well in the transfer learning context. As shown in Sect. 3, the optimization problem remains convex and can be directly solved using standard quadratic optimization tools. Introducing a MMD-like constraint is a heuristic which is equivalent to projecting the data onto a subspace where marginal distributions of observations are expected to become similar for both source and target data. Therefore, the discriminant function found by SVM for source data might perform well for target data. The experimental results prove the effectiveness of our idea.

Another way to make the marginal distributions of source and target observations similar is to first define a subspace adapted to the distributions (obtained, for example, by Principal Component Analysis) of source and target data respectively and then, to consider that both source and target data share the same coordinate system. To deal with potential non linear transformations between source and target data, *Kernel Principal Component Analysis (KPCA)* will be used here. We select a l-dimensional common subspace in such a way that MMD is minimized. This method is called KPCA alignment. It can be further improved by considering an additional transformation for which the parameters result from reducing the value of the previously obtained MMD.

Experimental results show the accuracy improvement of *KPCA* alignment, compared to that of the *MMD-like-constraint* alignment.

This paper is organized as follows: in Sect. 2, we give a short summary of related work; then we present our *MMD-like constrained SVM* method in Sect. 3 together with the optimization solution to the problem (in Sect. 4); the extension to *KPCA alignment* follows in Sect. 5; we prove the effectiveness of the proposed methods on synthetic and real data sets in Sect. 6. Finally, we conclude this paper and suggest perspectives to our work.

2   Related Work

In this section, we first review the common existing transductive transfer learning methods, especially domain adaptation methods. Then we present SVM based transfer learning and MMD based transfer learning, which is closely related to the first part of our work. Some dimension reduction transfer learning methods are presented, which are related to the second part of our work (KPCA alignment). For more transfer learning methods, recent advances of transfer learning, readers are referred to [11,19,20], etc. We start by a brief review of general transductive transfer learning.

According to [19], transductive transfer learning methods solve the problems where source and target share the same label space while differ in feature space. In this category of transfer learning, there are sample selection bias, co-variate shift and domain adaptation. Sample selection bias and co-variate shift deal with situations where only marginal distributions of source and target are different, conditional distributions are required to be the same; domain adaptation refers to the situations where marginal distributions and/or conditional distributions can be different between source and target. When target labels are unavailable, authors proposed instance reweighting strategies: typically [10,23], both of which solve sample selection bias and co-variate shift problems; others have also proposed domain adaptation strategies, for example GFK [7], TCA [18], JDA [15], etc.

SVM based transfer learning have been applied in many applications, from information retrieval to pedestrian detection. In general, authors modify the standard SVM to adapt to the transfer learning context. To the best of our knowledge, some authors define $w_{target} = w_{common} + w_{specific}$ and the two latter parameters are found from sources [36]; some modify the penalty by multiplying a reweighting factor to the penalty term of standard SVM [13]; some add an extra regularization term to standard SVM to control transfer [9,31], etc. So far, we have not seen SVM based transfer learning methods that have an extra constraint to control transfer, which is one of the novelties of our work.

MMD is an efficient measure of similarity between distributions and is widely used in transfer learning methods. MMD based transfer learning integrates MMD into traditional machine learning methods, aiming at controlling the transfer by MMD. However, most of the MMD-based transfer learning methods contribute to using MMD as an extra regularization term, which will balance the specific classification performance and the transfer. Some others use MMD as a pre-selection criterion to eliminate unrelated source domains. Interested readers are referred to [18,22,24,33,35], etc.

We now introduce related works of KPCA transfer learning. In general, KPCA is a traditional method for dimension reduction and usually takes the role of preprocessing in data mining. There are works related to dimension reduction which have contributed to solving transfer learning problems: [27] looks for a way of dimension reduction, after which the target data keeps their discriminative characteristics while it can also benefit from the advantage of transfer learning; [17] performs dimension reduction in the learning process of a universal kernel for transfer problem; [32,34] use the linear discriminant analysis while adapting the scatter matrix in different ways to different transfer learning contexts; [1] tries to find a common matching subspace while taking into consideration, as well, the conditional probability density function of the subspace-source-data, etc. To the best of our knowledge, KPCA has not been used to deal with transfer learning problems directly. Compared to the methods listed previously, applying directly KPCA to transfer learning is simple.

3   Presentation of the MMD Constrained SVM Method

In this section, we briefly present an extended version of our MMD constrained SVM transfer learning method which has been presented in [2]. We start by presenting some necessary fundamentals followed by the details of our approach.

3.1   Review of Basic Theoretical Foundations

Maximum Mean Discrepancy. Introduced in [5], Maximum Mean Discrepancy (MMD) is a non-parametric *distance* between two probability distributions. It measures the maximum distance between the expected values of these distributions (any distribution p and any distribution q) w.r.t any transformation ($f : x \rightarrow f(x)$, where x is a random variable drawn from the distribution)

$$MMD[\mathcal{F}, p, q] = \sup_{f \in \mathcal{F}} (\mathbf{E}_p[f(x)] - \mathbf{E}_q[f(y)])$$

In [3], from the Theorem on MMD, we can conclude that distributions p and q are equal iff $MMD = 0$. Then, using kernel embedding of distributions, Smola [28] and Gretton et al. [8] have shown that MMD can be easily evaluated in a Reproducing Kernel Hilbert Space (RKHS). Accordingly, MMD can be expressed as $MMD = ||\mu_p - \mu_q||_{\mathcal{H}}$, where $\mathcal{H}$ represents a RKHS, $\mu_{\{p,q\}}$ stands for $\mathbf{E}_{\{p,q\}}[k(x,.)]$ and $k(x,.)$ is the representation of x in the RKHS, which is equivalent to any transformation function f because of the nature of kernel. In the previous expression, the kernel must be *universal*[1]. The demonstration is shown as follows:

---

[1] A *universal* kernel is necessarily characteristic, while the reverse is not true. For more details see [29,30].

$$MMD^2[\mathcal{F}, p, q] = [\sup_{\|f\|_{\mathcal{H}} \le 1} (\mathbf{E}_p[f(x)] - \mathbf{E}_q[f(y)])]^2$$

$$= [\sup_{\|f\|_{\mathcal{H}} \le 1} (\mathbf{E}_p[<\phi(x), f>_{\mathcal{H}}] - \mathbf{E}_q[<\phi(y), f>_{\mathcal{H}}])]^2$$

$$= [\sup_{\|f\|_{\mathcal{H}} \le 1} <\mu_p - \mu_q, f>_{\mathcal{H}}]^2$$

$$= \|\mu_p - \mu_q\|_{\mathcal{H}}^2$$

Then using the kernel trick, the squared MMD can be further developed as follows:

$$MMD^2[\mathcal{F}, p, q] = \|\mu_p - \mu_q\|_{\mathcal{H}}^2$$

$$= \mathbf{E}_{p,p}[k(x, x')] - 2\mathbf{E}_{p,q}[k(x, y)] + \mathbf{E}_{q,q}[k(y, y')]$$

Here, x and x' are independent observations drawn from distribution p, y and y' are independent observations from distribution q, k designates a *universal* kernel function.

Theorem 1 *(Steinwart [30] and Smola [28])*.
$MMD[\mathcal{F}, p, q] = 0$ *iff* $p = q$ *when* $\mathcal{F} = \{f : \|f\|_{\mathcal{H}} \le 1\}$ *provided that* $\mathcal{H}$ *is universal.*

To make kernelized MMD calculable, an unbiased estimation is proposed in [26]:

$$\widehat{MMD}_u^2[\mathcal{F}, X, Y] = \frac{1}{m(m-1)} \sum_{i=1}^{m} \sum_{j \ne i}^{m} k(x_i, x_j)$$

$$+ \frac{1}{n(n-1)} \sum_{i=1}^{n} \sum_{j \ne i}^{n} k(y_i, y_j) - \frac{2}{nm} \sum_{i=1}^{m} \sum_{j=1}^{n} k(x_i, y_j) \tag{1}$$

where $x_i, i = 1, \ldots, m$ and $y_i, i = 1, \ldots, n$ are iid examples drawn from p and q respectively.

SVM. As our work is based on SVM, we remind the primal optimization problem for soft margin SVMs.

$$\min_{w, \varepsilon, b} \frac{1}{2}\|w\|^2 + C \sum_{i=1}^{n} \varepsilon_i$$

$$\text{s.t. } \varepsilon_i \ge 0, \quad \forall i = 1, \ldots, n$$

$$y_i(<w, \phi(x_i)> + b) \ge 1 - \varepsilon_i, \quad \forall i = 1, \ldots, n$$

where w is the vector normal to the hyperplane defined in the RKHS, ε_i is the error term associated to observation i, C is the trade-off parameter between the margin term and classification error, $\phi(x_i)$ is the kernel representation of x_i, y_i is the label of x_i and b is the bias.

The objective of standard SVM is to find the best w which defines the classifier $f = sign(<w, \phi(x)> + b)$ that maximally separates positive and negative classes. The use of $\phi(.)$ makes nonlinear classifiers possible, which corresponds to most of the real classification tasks.

3.2   MMD-Like Constrained SVM Transfer Learning

In this subsection, we briefly present the heuristic proposed in our previous paper [2].

With the assumption (in Sect. 1) satisfied, the objective of this first heuristic is to find a classifier that maximally separates source classes while remains effective for the target classification task. Therefore, we modify the standard SVM by adding an extra constraint, whose aim is to force the hyperplane to lie in a subspace of the RKHS where source and target data are made similar. The similarity is measured by the estimated MMD defined in Eq. 1. The general formulation of the proposed approach is as follows:

$$\min_{w,\varepsilon,b} \frac{1}{2}||w||^2 + C \sum_{i=1}^{n} \varepsilon_i$$
$$\text{s.t.} <\mu_{X_s} - \mu_{X_t}, w>_{\mathcal{H}} = 0 \tag{2}$$
$$\varepsilon_i \geq 0, \quad \forall i = 1, \ldots, n$$
$$y_i(<w, \phi(x_i)> + b) \geq 1 - \varepsilon_i, \quad \forall i = 1, \ldots, n$$

where the second line is the extra constraint and the other parts are the same as standard SVM; μ_{X_s} (μ_{X_t}) is the sample mean of source (target) data in $\mathcal{H}$ and can be estimated by $\mu_{X_s} = \frac{1}{n_s} \sum \phi(X_s)$ ($\mu_{X_t} = \frac{1}{n_t} \sum \phi(X_t)$).

To understand the role of the extra constraint, making $<\mu_{X_s} - \mu_{X_t}, w>_{\mathcal{H}} = 0$ corresponds to limiting the objective hyperplane (whose direction is determined by w) to lie in the subspace where $MMD = 0$. In this way, we expect that source and target data will be as similar as possible in this subspace. Accordingly, if the hyperplane can well classify source data, it might perform well on target data.

In this paper, we use a MMD-like constraint instead of a MMD-like regularization term (as in [22]). This allows to focus on the transfer ability rather than on performance on source data. If we compare our approach with that of Quanz and Huan [22], where there is an extra regularization parameter λ (whose value influences the trade-off between transfer effect and the classification performance on source data), with a finite value of λ, we might sacrifice the similarity between source and target data to achieve high performance only for source data.

Moreover, as shown in Sect. 4, our heuristic avoids the calculation of the inverse of a matrix which leads to inefficiency and inaccuracy during the optimization process. In [22], Quanz and Huan proposed to alleviate this problem by approximating the original matrix by generalized singular value decomposition, but the calculation of the inverse of a matrix was unavoidable.

4   Dual Form of the Optimization Problem

We first use the representer theorem [25] to solve problem (2), so w can be expressed as:

$$w = \sum_{k=1}^{n_s} \beta_k^s \phi(x_k^s) + \sum_{l=1}^{n_t} \beta_l^t \phi(x_l^t) \tag{3}$$

where β_k^s and β_l^t are the parameters to be determined.

Then, the extra constraint can be expressed in terms of β:

$$<\mu_{X_s} - \mu_{X_t}, w>_{\mathcal{H}} = \frac{1}{n_s}\sum_{k=1}^{n_s}\beta_k^s\sum_{i=1}^{n_s}<\phi(x_i),\phi(x_k)>_{\mathcal{H}} - \frac{1}{n_t}\sum_{k=1}^{n_s}\beta_k^s\sum_{j=1}^{n_t}<\phi(x_j),\phi(x_k)>_{\mathcal{H}}$$

$$+ \frac{1}{n_s}\sum_{l=1}^{n_t}\beta_l^t\sum_{i=1}^{n_s}<\phi(x_i),\phi(x_l)>_{\mathcal{H}} - \frac{1}{n_t}\sum_{l=1}^{n_t}\beta_l^t\sum_{j=1}^{n_t}<\phi(x_j),\phi(x_l)>_{\mathcal{H}}$$

$$= (\begin{bmatrix} K_{SS} & K_{TS} \\ K_{ST} & K_{TT} \end{bmatrix}[\underbrace{\frac{1}{n_s},\ldots,\frac{1}{n_s}}_{n_s},\underbrace{-\frac{1}{n_t},\ldots,-\frac{1}{n_t}}_{n_t}]^T)^T[\beta^s,\beta^t]^T$$

$$= (K\tilde{1})^T\beta$$

The new expression of problem (2) in terms of β becomes:

$$\min_{\beta,\varepsilon,b} \frac{1}{2}\beta^T K\beta + C\sum_{i=1}^{n}\varepsilon_i$$

$$\text{s.t.}(K\tilde{1})^T\beta = 0 \qquad\qquad (4)$$

$$\varepsilon_i \geq 0, \quad \forall i = 1,\ldots,n$$

$$y_i(\beta^T<\phi(X),\phi(x_i)> + b) \geq 1 - \varepsilon_i, \quad \forall i = 1,\ldots,n$$

Using Lagrangian optimization, we obtain the dual form:

$$\max_{\mu,\eta} \sum_{i=1}^{n_s}\mu_i - \frac{1}{2}(\sum_{i=1}^{n_s}\mu_iy_iK_{.i})^TK^{-1}(\sum_{j=1}^{n_s}\mu_jy_jK_{.j}) - \frac{1}{2}\eta^2\tilde{1}^TK^T\tilde{1} - \eta(\sum_{i=1}^{n_s}\mu_iy_iK_{.i})^T\tilde{1}$$

$$\text{s.t. } 0 \leq \mu_i \leq C \text{ and } \sum_{i=1}^{n_s}\mu_iy_i = 0$$

where $K_{.i} = <\phi(X),\phi(x_i)>_{\mathcal{H}}$ and X represents the ensemble of X_s and X_t; x_i is a single observation either from X_s or X_t.

If we fix μ and consider the optimization only with regards to η, we can find the optimal value of η in terms of μ, $\eta = -\frac{(\sum_{i=1}^{n_s}\mu_iy_iK_{.i})^T\tilde{1}}{\tilde{1}^TK^T\tilde{1}}$. We now obtain the final dual form, which is still quadratic w.r.t μ:

$$\max_{\mu} \sum_{i=1}^{n_s}\mu_i - \frac{1}{2}(\sum_{i=1}^{n_s}\mu_iy_iK_{.i})^T(K^{-1} - \frac{\tilde{1}\tilde{1}^T}{\tilde{1}^TK\tilde{1}})(\sum_{j=1}^{n_s}\mu_jy_jK_{.j})$$

$$\text{s.t. } 0 \leq \mu_i \leq C \text{ and } \sum_{i=1}^{n_s}\mu_iy_i = 0.$$

With some further calculation, we have:

$$\max_{\gamma} \gamma^TY - \frac{1}{2}\gamma^T(K_{SS} - \frac{K_{S.}\tilde{1}\tilde{1}^TK_{S.}^T}{\tilde{1}^TK^T\tilde{1}})\gamma$$

$$\text{s.t. } \sum_{i=1}^{n_s}\gamma_i = 0 \text{ and } \min(0,Cy_i) \leq \gamma_i \leq \max(0,Cy_i), \forall i = 1,\ldots,n_s.$$

where $\gamma_i = \mu_i y_i$ and $K_{S.} = \sum_{i=1}^{n_s} K_{i.}$. The matrix $K_{SS} - \frac{K_{S.}\widetilde{11}^T K_S^T}{\widetilde{1}^T K^T \widetilde{1}}$ is the matrix of inner products (in the subspace orthogonal to w) of source data. Following the demonstration in [21], we can conclude that if $\mathcal{H}$ is a RKHS on X and $\mathcal{H}_0 \in \mathcal{H}$ is a closed subspace, then $\mathcal{H}_0$ is also a RKHS on X. Therefore, as the matrix $K_{new} = K_{SS} - \frac{K_{S.}\widetilde{11}^T K_S^T}{\widetilde{1}^T K^T \widetilde{1}}$ is the new Gram matrix corresponding to the projected kernel, K_{new} is positive semi-definite.

5   Extension to KPCA Alignment

The principle of MMD-like constrained transfer learning method is to match source and target data in a subspace of a RKHS so that the classifier for source can work well on target data. Although $MMD = 0$ guarantees the equality of two distributions, $<\mu_{X_s} - \mu_{X_t}, w>_{\mathcal{H}} = 0$ is no longer the MMD defined in Theorem 1. There is no guarantee that any transformation of the initial data, leading to a zero value of the *MMD-like constraint*, ensures that the transformed data distributions are similar. Although several experiments have proved the efficiency of our method (in Sect. 6), theoretical error bounds are to be developed. We now propose an alternative, based again on the idea that if marginal distributions of observations from source and target data become similar, the classifier trained on source data might perform well on target data. Our alternative to solving the domain adaptation problem is based on Kernel Principal Component Analysis (KPCA). This contribution briefly appeared in the perspectives of our previous paper [2] and is now presented in details. In this section, we start by presenting the fundamentals of PCA and KPCA, after which the details of our method will be shown.

5.1   A Brief Review of PCA and KPCA

Before introducing the details of KPCA, we first review the fundamentals of Principal Component Analysis (PCA). PCA is an unsupervised statistical procedure that linearly transforms data to a new orthonormal coordinate system so that the greatest variance lies on the first coordinate, the second greatest variance on the second coordinate and so on. A principal component designates each coordinate.

Let X be the matrix $(n \times m)$ of observations, the objective is to find the vector u so that Xu extract the most important part of data information, here measured by the variance of Xu. We begin by determining the first principal component defined by the vector u that captures the largest variance of Xu. We first center the data: $X_c = X - E(X)$ and the problem is defined as follows:

$$\arg\max_u ||X_c u||^2 = \arg\max_u u^T X_c^T X_c u \quad \text{s.t. } u^T u = 1 \tag{5}$$

The constraint $u^T u = 1$ guarantees that the optimization problem is not ill-posed. Introducing the Lagrangian, we have:

$$\mathcal{L} = u^T X_c^T X_c u - \lambda(u^T u - 1)$$

Therefore,

$$\frac{\partial \mathcal{L}}{\partial u} = 2X_c^T X_c u - 2\lambda u = 0 \Leftrightarrow X_c^T X_c u = \lambda u$$

From the above expression, we can find that the Lagrange parameter λ can be considered as an eigenvalue of $X_c^T X_c$ and the first principal direction u is the eigenvector that corresponds to the largest eigenvalue λ. With a similar reasoning, we understand that the lth principal component is the eigenvector that corresponds to the lth largest eigenvalue of the covariance matrix $X_c^T X_c$.

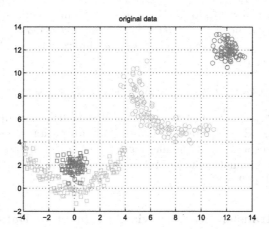

(a) Original data. *Squares* represent the source and *circles* represent the target.

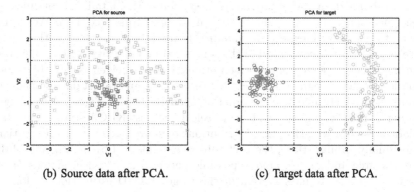

(b) Source data after PCA.   (c) Target data after PCA.

Fig. 1. Illustration of permutation of abscissa and ordinate after PCA. In both (b) and (c), abscissa corresponds to the eigenvector V1 associated to the largest eigenvalue, while ordinate corresponds to eigenvector V2 associated to the second largest eigenvalue.

KPCA is a kernelized version of PCA that extracts the principal components in a RKHS. After transformation of the data into the RKHS, the covariance matrix of the data becomes: $C_K = \frac{1}{n}\sum_{i=1}^{n}\phi_c(x_i)\phi_c(x_i)^T$, where $\phi_c(x_i) = \phi(x_i) - \frac{1}{n}\sum_{i=1}^{n}\phi(x_i)$; $\phi(X)$ is the kernel transformation of original data X; $\phi_c(x_i), i = 1, \ldots, n$ centers all the data in the high-dimensional space spanned by $\phi(.)$, corresponding to X_c in (5). As KPCA manipulates data in a RKHS, eigenvectors $(\bar{V})$ should lie in the space spanned by $\phi(x_i)$. So we have $\bar{V} = \sum_{i=1}^{n}\alpha_i\phi(x_i)$. We now have:

$$\tilde{\lambda}\bar{V} = C_K\bar{V} \Leftrightarrow n\tilde{\lambda}\alpha = M\alpha$$

where $M = (I_n - \frac{1}{n}1_n1_n^T)K(I_n - \frac{1}{n}1_n1_n^T)$, $K = <\phi(X),\phi(X)>_{\mathcal{H}}$, I_n is the identity matrix of size $n \times n$ and 1_n is the $n \times 1$ column vector with all elements equaling 1. We next normalize $\bar{V}$ ($\tilde{V}\tilde{V}^T = 1$) and get $\tilde{\alpha}M\tilde{\alpha} = 1$, which is equivalent to $n\tilde{\lambda}\tilde{\alpha}^T\tilde{\alpha} = 1$. Then, expressing the coordinates $(\tilde{V}\phi_c(Z))$ of any data set Z after KPCA, we obtain:

$$\tilde{V}\phi_c(Z) = \tilde{\alpha}K_c \text{ where } K_c(i,j) = <\phi_c(x_i),\phi_c(z_j)>_{\mathcal{H}}$$

In general, a subspace associated to a few l first principal components ($l \ll m$) is enough for a good data representation. The information represented by other components corresponding to smaller eigenvalues is usually associated to noise. In Sect. 6, on the simulated and real data sets considered in this paper, l can be selected as small as 2 or 3.

5.2   KPCA Transfer Learning via Alignment of Data Representations

The general principle of KPCA based transfer learning is the following. We apply KPCA to source data and obtain a representation of the data in a (non linearly) transformed space, in a coordinate system adapted to source data. We do the same for target data. Then the coordinates of target data (obtained in their coordinate system) are directly used in the source data coordinate system. This performs alignment between both representations. However, eigenvectors are defined up to the constant ± 1 and differences between source and target data can change the ranking of eigenvalues (so the order of eigenvectors, see Fig. 1). Accordingly, when performing alignment, we have to select the best KPCA subspace for target data in such a way that the distributions of source and target observations are maximally similar. Here, the best subspace is selected by minimizing MMD between source and target, among possible permutations and inversions of the elements of the coordinate system of the target. MMD estimation is done in a new RKHS, as presented in Theorem 1 in Sect. 1. The subspace dimension is determined, as usual, by selecting the number of axes that allow to preserve a reasonable percentage of the initial variance of source data.

Finally, after the coordinate system is optimized for target data to align, as well as possible, with source data, we can furthermore reduce the residual MMD by considering a linear transformation whose parameters (transformation matrix A, translation vector b) result, here again, from the minimization of this residual MMD w.r.t A and b.

In the remaining of this paper, KPCA transfer learning without linear transformation is called KPCA-TL and KPCA-LT-TL is the method using the additional linear transformation.

6   Experiments

In this section, we first take the similar synthetic and real data sets from the paper [2], because our first results contribute to comparing our first method with others from literature. Then, we apply our second method, KPCA based transfer learning, on these data sets. Furthermore, we also compare both methods with other transfer learning benchmark methods on new real data sets. Finally, we analyze the results and prove the efficiency of our methods.

6.1   Data Sets

We take the well-known banana-orange data set as our synthetic data set. We fix a source data set and generate a target data by drawing samples from a translated and distorted version of the source data distribution. For binary classification task, we can attribute positive labels to the banana while negative labels to the orange. To form a domain adaptation problem, we suppose that the label information of source is known while it is unknown for target data (see an example in Fig. 3(a)).

For real data sets, we first use the USPS real data set, a handwritten digits data set. There are training and testing subsets, both containing the images (16×16 pixels) of handwritten digits 0 to 9. We use the training subset as our source data while the testing subset is our target data. We suppose that there is no label information for target data. As in [33], the objective is to separate digit 4 and digit 7 as the source task and to separate digit 4 and digit 9 as the target task (USPS-Task 1). Other similar transfer learning data set can be formed, for example, the classification of digits 3 and 6 (source) to help classification of digits 3 and 8 (target). Generally, we take advantage of an easier task to help the classification of a harder task (see Fig. 2, where a t-SNE plot [16] illustrates this case).

Some other data sets from the UCI data repository[2] are also used here, namely IRIS and SEED. For IRIS, we know that there are 3 classes, easily separated. For the source, we take iris-setosa and iris-versicolor as source positive and source negative class, while for the target, we consider iris-versicolor as negative class and iris-virginica as positive class (IRIS-Task 1), respectively. We can also form another transfer learning task by using iris-setosa and iris-virginica as source while iris-versicolor and iris-virginica as target (IRIS-Task 2). Similarly for SEED data set (SEED-Task 1: source data are the Canadian wheat variety and the Rosa wheat variety while target are the Canadian wheat variety and the Kama wheat variety. SEED-Task 2: the Canadian wheat variety and the Rosa

---

[2] http://archive.ics.uci.edu/ml/.

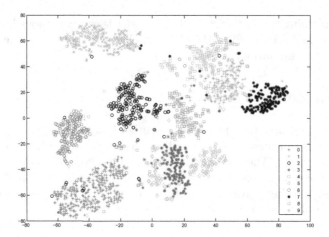

Fig. 2. t-SNE plot of USPS testing data. Group 0 is represented by red *plus*; 1 by blue *x-marks*; 2 by black *circles*; 3 by cyan *stars*; 4 by magenta *squares*; 5 by magenta *diamonds*; 6 by red *circles*; 7 by black *stars*; 8 by blue *squares*; 9 by blue *plus*. (Color figure online)

wheat variety are source data while the Kama wheat variety and the Rosa wheat variety are the target).

In Table 1, we compare the results obtained with our methods and that with LM (the method proposed in [22]). In [22], LM has been proved superior to some other related transfer learning methods (T-SVM in [12], CDSC in [14], LWE in [6]), so we omit here the comparison to these methods. Other domain adaptation methods are also included in the comparison, even though some of the methods have little relatedness with ours.

From this general comparison, we show readers that, for some data sets, our methods give similar or even better results. Included methods are TCA [18], GFK [7], JDA [15]. Standard SVM is also compared to show the usefulness of transfer learning (or cases when negative transfer happens).

6.2   Experimental Results and Analysis

For the banana-orange data set (Fig. 3(a)), we show the classification result obtained comparing SVMMMD, LM and KPCA-TL (or KPCA-LT-TL [3]) (see Fig. 3(a), (c) and (f)). Standard SVM is not compared in this figure, because standard SVM can not well classify this data set. Figure 3 presents the classification results on banana-orange data set for a representative random generation, including data sets, discriminant functions for source and target data, decision surfaces. For KPCA transfer learning methods, we present the classifier in the RKHS (Fig. 3(e)) and the classification results in the original space (Fig. 3(f)), as there is no need for pre-image to get the classification result.

---

[3] Here, KPCA-TL and KPCA-LT-TL lead to the same results.

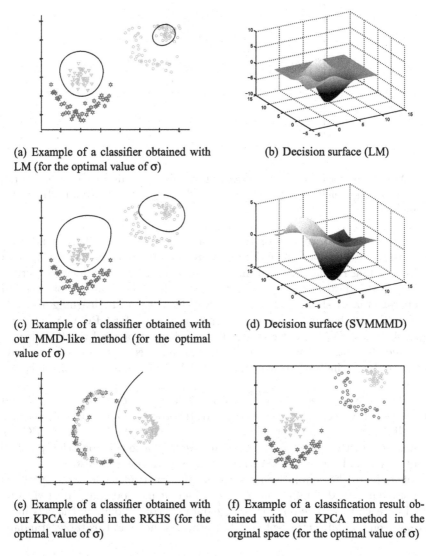

(a) Example of a classifier obtained with LM (for the optimal value of σ)

(b) Decision surface (LM)

(c) Example of a classifier obtained with our MMD-like method (for the optimal value of σ)

(d) Decision surface (SVMMMD)

(e) Example of a classifier obtained with our KPCA method in the RKHS (for the optimal value of σ)

(f) Example of a classification result obtained with our KPCA method in the orginal space (for the optimal value of σ)

Fig. 3. Results obtained on the banana-orange data set. In (a) and (c), *triangles* and *stars* represent the labeled source data while *circle* symbols are the unlabeled target data. In (d) and (b), decision surfaces are plotted as functions of the input space coordinates. Thresholding these surfaces at 0 level gives the decision curves corresponding to the classifiers shown in (c) and (a), respectively. (e) and (f) represent the classification results of our KPCA transfer learning methods (results are identical for both methods). In (e), *triangles* and *stars* are source data while *circles* denote target data, colors represent classes: red for orange and blue for banana. In (f), source is represented by red *triangles* and blue *stars* while target data is represented by *circles*; colors of circles represent the classification result (before KPCA transfer learning, target are unlabeled). (Color figure online)

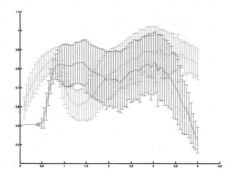

Fig. 4. Average performance (good classification rate) ±1 s.d. as a function of the gaussian kernel parameter. Red line: our method. Blue line: LM. (from [2]). (Color figure online)

For an average performance on banana-orange data set, 50 different independently generated banana-orange data sets are used to compare SVMMMD and its corresponding regularization-term-based approach LM. The average performances (±1 standard deviation) of both approaches as a function of gaussian kernel parameter are shown in Fig. 4. However, KPCA transfer learning methods are not compared, because there are 2 kernel parameters for KPCA (σ_{KPCA} and σ_{MMD}). Standard SVM is not compared either, because a good result on target data cannot be obtained without transfer learning (see Table 1). We conclude that our method achieves better results than LM for a wider range of the kernel parameter.

For the USPS data set, if we use digit 3 and digit 6 as source while digit 3 and digit 8 as target, we can find that our SVMMMD provides higher performance for almost all the kernel parameter values considered.(referred to [2] Fig. 6) KPCA is not compared for the same reason as before. We have also tried on other possible pairs of source and target, SVMMMD generally performs the best.

Table 1 shows the comparison results of different transfer learning methods and of standard SVM. The parameters for different methods are adjusted to be optimal for each task. The banana-orange data set corresponds to the classification results on the data represented in Fig. 3(a). As shown in this table, standard

Table 1. Good classification results on different datasets for different transfer learning methods.

	SVM	TCA	JDA	GFK	LM	SVM MMD	KPCA-TL	KPCA-LT-TL
USPS-Task 1	0.6976	0.7347	0.8329	0.7560	0.7294	0.9496	0.8117	0.8143
SEED-Task 1	0.8000	0.7286	0.7786	0.7857	0.8929	0.9071	0.9357	0.9214
SEED-Task 2	0.7000	0.8214	0.7714	0.7214	0.8571	0.9357	0.9214	0.9143
IRIS-Task 1	0.5000	0.8800	0.5000	0.5000	0.7000	0.8700	**0.9000**[a]	**0.9300**[a]
IRIS-Task 2	0.5000	0.5000	0.5100	0.5000	0.7300	0.9500	0.9300	0.9700
Banana-Orange	0.5062	0.7654	0.6914	0.4938	0.9012	0.9877	1.0000	1.0000

[a]Results obtained after an inversion of labels, read the comment in the text.

SVM can not solve the problem in 3 out of 6 tasks. For these 3 cases, standard SVM will predict all the target data in the same class, which is obviously wrong. Our methods achieve the best results in almost all tasks.

However, for KPCA transfer learning methods (KPCA-TL and KPCA-LT-TL), positive and negative groups of source and target data might be mistaken after KPCA, leading to an inversion of the labels (for example, in IRIS-Task 1, the labels obtained should be inverted, what we have done here). This confusion can be easily corrected if a few labeled target data observations are available. GFK does not work well, perhaps because there is no smooth interpolation between source and target. JDA works relatively well as it utilizes the stability condition for clustering in solving transfer learning problems, but our experiments show that, in some cases, JDA does not converge and perhaps sometimes gets trapped in local minima. TCA is similar to LM and seems to be outperformed mostly because they might sacrifice their target classification performance to source classification. On the data sets considered in this study, our methods compare favorably with the alternatives considered.

7 Conclusion and Future Directions

In this paper, we propose new methods to solve the domain adaptation problem when no labeled target data is available. We suppose that there exists a smooth nonlinear transformation of the target data distribution that makes it similar to that of source data. We also suppose that, after this transformation, the conditional probability distributions of labels remain similar for source and target. The first approach is to perform a projection of source and target data onto a subspace of a RKHS where source and target data distributions are expected to be similar. To do so, we select the subspace which ensures nullity of a *Maximum Mean Discrepancy* based criterion. As source and target data become similar, the SVM classifier trained on source data performs well on target data. We have shown that this additional constraint on the primal optimization problem does not modify the nature of the dual problem so that standard quadratic programming tools can be used. Following the same principal idea, we extend our method to Kernel based Principal Component Analysis transfer learning methods, which in most cases improve the classification performance. We have applied our methods on synthetic and real data sets and have shown that our results favorably compare with other transfer learning methods.

As short term developments, we must propose a method to automatically determine an adequate value of the gaussian kernel parameters used. We also have to investigate the deduction of error bounds, which appears to be a challenging task.

References

1. Blöbaum, P., Schulz, A., Hammer, B.: Unsupervised dimensionality reduction for transfer learning. In: Proceedings of the 23rd European Symposium on Artificial Neural Networks, Computational Intelligence and Machine Learning (2015)
2. Chen, X., Lengellé, R.: Domain adaptation transfer learning by SVM suhject to a maximum-mean-discrepancy-like constraint. In: Proceedings of the 6th International Conference on Pattern Recognition Applications and Methods, ICPRAM 2017 (2017)
3. Dudley, R.M.: A course on empirical processes. In: Hennequin, P.L. (ed.) École d'Été de Probabilités de Saint-Flour XII - 1982. LNM, vol. 1097, pp. 1–142. Springer, Heidelberg (1984). https://doi.org/10.1007/BFb0099432
4. Dudley, R.M.: Real Analysis and Probability, vol. 74. Cambridge University Press, Cambridge (2002)
5. Fortet, R., Mourier, E.: Convergence de la répartition empirique vers la réparation théorique. Ann. Scient. École Norm. Sup., 266–285 (1953)
6. Gao, J., Fan, W., Jiang, J., Han, J.: Knowledge transfer via multiple model local structure mapping. In: Proceedings of the 14th ACM SIGKDD International Conference on Knowledge Discovery and Data Mining, pp. 283–291. ACM (2008)
7. Gong, B., Shi, Y., Sha, F., Grauman, K.: Geodesic flow kernel for unsupervised domain adaptation. In: 2012 IEEE Conference on Computer Vision and Pattern Recognition (CVPR), pp. 2066–2073. IEEE (2012)
8. Gretton, A., Borgwardt, K.M., Rasch, M.J., Schölkopf, B., Smola, A.: A kernel two-sample test. J. Mach. Learn. Res. **13**, 723–773 (2012)
9. Huang, C.-H., Yeh, Y.-R., Wang, Y.-C.F.: Recognizing actions across cameras by exploring the correlated subspace. In: Fusiello, A., Murino, V., Cucchiara, R. (eds.) ECCV 2012. LNCS, vol. 7583, pp. 342–351. Springer, Heidelberg (2012)
10. Huang, J., Gretton, A., Borgwardt, K.M., Schölkopf, B., Smola, A.J.: Correcting sample selection bias by unlabeled data. In: Advances in Neural Information Processing Systems, pp. 601–608 (2006)
11. Jiang, J.: A literature survey on domain adaptation of statistical classifiers (2008). http://sifaka.cs.uiuc.edu/jiang4/domainadaptation/survey
12. Joachims, T.: Transductive inference for text classification using support vector machines. In: ICML, vol. 99, pp. 200–209 (1999)
13. Liang, F., Tang, S., Zhang, Y., Zuoxin, X., Li, J.: Pedestrian detection based on sparse coding and transfer learning. Mach. Vis. Appl. **25**(7), 1697–1709 (2014)
14. Ling, X., Dai, W., Xue, G.-R., Yang, Q., Yu, Y.: Spectral domain-transfer learning. In: Proceedings of the 14th ACM SIGKDD International Conference on Knowledge Discovery and Data Mining, pp. 488–496. ACM (2008)
15. Long, M., Wang, J., Ding, G., Sun, J., Yu, P.S.: Transfer feature learning with joint distribution adaptation. In: Proceedings of the IEEE International Conference on Computer Vision, pp. 2200–2207 (2013)
16. van der Maaten, L., Hinton, G.: Visualizing data using t-SNE. J. Mach. Learn. Res. **9**(Nov), 2579–2605 (2008)
17. Pan, S.J., Kwok, J.T., Yang, Q.: Transfer learning via dimensionality reduction. In: AAAI, vol. 8, pp. 677–682 (2008)
18. Pan, S.J., Tsang, I.W., Kwok, J.T., Yang, Q.: Domain adaptation via transfer component analysis. IEEE Trans. Neural Netw. **22**(2), 199–210 (2011)
19. Pan, S.J., Yang, Q.: A survey on transfer learning. IEEE Trans. Knowl. Data Eng. **22**(10), 1345–1359 (2010)

20. Patel, V.M., Gopalan, R., Li, R., Chellappa, R.: Visual domain adaptation: a survey of recent advances. IEEE Signal Process. Mag. **32**(3), 53–69 (2015)
21. Paulsen, V.I.: An Introduction to the Theory of Reproducing Kernel Hilbert Spaces. Cambridge University Press, Cambridge (2009)
22. Quanz, B., Huan, J.: Large margin transductive transfer learning. In: Proceedings of the 18th ACM Conference on Information and Knowledge Management, pp. 1327–1336. ACM (2009)
23. Quionero-Candela, J., Sugiyama, M., Schwaighofer, A., Lawrence, N.D.: Dataset Shift in Machine Learning. The MIT Press, Cambridge (2009)
24. Ren, J., Liang, Z., Hu, S.: Multiple kernel learning improved by MMD. In: Cao, L., Zhong, J., Feng, Y. (eds.) ADMA 2010. LNCS (LNAI), vol. 6441, pp. 63–74. Springer, Heidelberg (2010). https://doi.org/10.1007/978-3-642-17313-4_7
25. Schölkopf, B., Herbrich, R., Smola, A.J.: A generalized representer theorem. In: Helmbold, D., Williamson, B. (eds.) COLT 2001. LNCS (LNAI), vol. 2111, pp. 416–426. Springer, Heidelberg (2001). https://doi.org/10.1007/3-540-44581-1_27
26. Serfling, R.J.: Approximation Theorems of Mathematical Statistics, vol. 162. Wiley, Hoboken (2009)
27. Si, S., Tao, D., Geng, B.: Bregman divergence-based regularization for transfer subspace learning. IEEE Trans. Knowl. Data Eng. **22**(7), 929–942 (2010)
28. Smola, A.: Maximum mean discrepancy. In: Proceedings of the 13th International Conference, ICONIP 2006, Hong Kong, China, 3–6 October 2006
29. Sriperumbudur, B.K., Gretton, A., Fukumizu, K., Schölkopf, B., Lanckriet, G.R.G.: Hilbert space embeddings and metrics on probability measures. J. Mach. Learn. Res. **11**(Apr), 1517–1561 (2010)
30. Steinwart, I.: On the influence of the kernel on the consistency of support vector machines. J. Mach. Learn. Res. **2**, 67–93 (2002)
31. Tan, Q., Deng, H., Yang, P.: Kernel mean matching with a large margin. In: Zhou, S., Zhang, S., Karypis, G. (eds.) ADMA 2012. LNCS (LNAI), vol. 7713, pp. 223–234. Springer, Heidelberg (2012). https://doi.org/10.1007/978-3-642-35527-1_19
32. Tu, W., Sun, S.: Transferable discriminative dimensionality reduction. In: 2011 23rd IEEE International Conference on Tools with Artificial Intelligence (ICTAI), pp. 865–868. IEEE (2011)
33. Uguroglu, S., Carbonell, J.: Feature Selection for transfer learning. In: Gunopulos, D., Hofmann, T., Malerba, D., Vazirgiannis, M. (eds.) ECML PKDD 2011. LNCS (LNAI), vol. 6913, pp. 430–442. Springer, Heidelberg (2011). https://doi.org/10.1007/978-3-642-23808-6_28
34. Wang, Z., Song, Y., Zhang, C.: Transferred dimensionality reduction. In: Daelemans, W., Goethals, B., Morik, K. (eds.) ECML PKDD 2008. LNCS (LNAI), vol. 5212, pp. 550–565. Springer, Heidelberg (2008). https://doi.org/10.1007/978-3-540-87481-2_36
35. Yang, S., Lin, M., Hou, C., Zhang, C., Yi, W.: A general framework for transfer sparse subspace learning. Neural Comput. Appl. **21**(7), 1801–1817 (2012)
36. Zhang, P., Zhu, X., Guo, L.: Mining data streams with labeled and unlabeled training examples. In: Ninth IEEE International Conference on Data Mining, ICDM 2009, pp. 627–636. IEEE (2009)

Optimal Linear Imputation
with a Convergence Guarantee

Yehezkel S. Resheff[1,2(✉)] and Daphna Weinshall[1]

[1] School of Computer Science and Engineering,
The Hebrew University of Jerusalem, Jerusalem, Israel
{heziresheff,daphna}@cs.huji.ac.il
[2] Edmond and Lily Safra Center for Brain Sciences,
The Hebrew University of Jerusalem, Jerusalem, Israel

Abstract. It is a common occurrence in the field of data science that real-world datasets, especially when they are high dimensional, contain missing entries. Since most machine learning, data analysis, and statistical methods are not able to handle missing values gracefully, these must be filled in prior to the application of these methods. It is no surprise therefore that there has been a long standing interest in methods for imputation of missing values. One recent, popular, and effective approach, the IRMI stepwise regression imputation method, models each feature as a linear combination of all other features. A linear regression model is then computed for each real-valued feature on the basis of all other features in the dataset, and subsequent predictions are used as imputation values. However, the proposed iterative formulation lacks a convergence guarantee. Here we propose a closely related method, stated as a single optimization problem, and a block coordinate-descent solution which is guaranteed to converge to a local minimum. Experiment results on both synthetic and benchmark datasets are comparable to the results of the IRMI method whenever it converges. However, while in the set of experiments described here IRMI often diverges, the performance of our method is shown to be markedly superior in comparison to other methods.

1 Introduction

The typical modus operandi in the field of data science evolves a wrangling stage where either the raw data or features computed on the basis of the raw data are organized in the form of a table. Indeed, the vast majority of data analysis, machine learning, and statistical methods rely on complete data [9], mostly structured in a tabular or relational form.

Since in real-world datasets more often than not some of the entries are missing, imputation is an important part of data preprocessing and cleansing [13,24].

Invited extension of [29] – presented at the 6th International Conference on Pattern Recognition Applications and Methods (ICPRAM2017).

Naturally, this topic has been of long-standing interest in many fields associated with data analysis.

As is often the case, simple and elegant linear methods with interpretable results have gained a special place in the heart of the field, and are used in practice whenever applicable. More advanced methods (see Sect. 3 for a brief review) are typically reserved for special cases.

The trivial option for many application domains is to discard complete records in which there are any missing values. Clearly this method is sub-optimal, for several reasons: first and foremost, when missing values are not missing at random [11,20], discarding these records may bias the resulting analysis [21] (consider for instance a classification task where many of the examples from one of the classes have some missing features. Discarding these entire examples will lead to a very unbalanced problem, with a potentially detrimental effect on the final results).

Other limitations include the needless loss of information when discarding entire records which may actually include valuable information for the down-stream task. Furthermore, when dealing with datasets with either a small number of records or a large number of features, omitting complete records when any feature value is missing may result in discarding a large proportion of the records (or in the extreme case – all of them), and insufficient data for the required analysis.

There are several method traditionally used to preform data imputation. These include procedures which impute missing values by replacing them with summary statistics such as the mean or median of the feature value across records [5,7]. While using such summary statistics may indeed provide satisfactory results when there is no other information present, this is however most-often not the case we are dealing with. Namely, for each missing feature value there are other non-missing values in the same record. For this reason it is likely (or in fact we assume) that other features contain information regarding the value of the missing feature, and imputation should therefore take into account known feature values in the same record. This is done by all subsequent methods.

The method of multiple imputation (see [31] for a detailed review) generates several sets of missing value imputations, drawn from the posterior distribution of the missing values under a given model, given the data. All down-stream processing is then performed on each copy of the imputed data, and the resulting multiple sets of model parameters are combined as a final step to produce a single result.

Multiple imputation methods are extremely useful in traditional statistical analysis and heavily utilized in analysis of public survey data. However, this may not be feasible in a machine learning and modern statistical setting, for several reasons. First, the run-time cost of performing the analysis on several copies of the full-data may be prohibitive. Second, being a model-based approach it depends heavily on the type and nature of the data, and can't be used as an out-of-the-box pre-processing step. More importantly though, while traditional model parameters may (for the most part) be combined between versions of the imputed data (regression coefficients for instance), many modern machine learning methods do not produce a representation that is straightforward to combine (consider the parameters of an Artificial Neural Network or a Random Forest for example[1]).

In [25], a method for imputation on the basis of a sequence of regression models is introduced. The method, popularized under the acronym MICE [1,36], uses a non-empty set of complete features (i.e. features with values which are known in all the records) as its base, and iteratively imputes one feature at a time on the basis of the completed features up to that point.

Since each step of this method produces a single complete feature, the number of iterations needed to impute the entire table is exactly the number of features that have a missing value in at least one record. The drawbacks of this method are twofold. First, there must be at least one complete feature to be used as the base (however if there is no complete feature then the feature with fewest missing values may be imputed using the a feature-wise summary statistic). More importantly though, the values imputed at the i-th step can only use a regression model that includes the features which were originally full or those imputed in the first $i - 1$ steps. Ideally, the regression model for each feature should be able to use all other feature values, thus not discarding any available information.

The IRMI method [34] goes one step further by again building a sequence of regression models for each feature, this time utilizing the values in all other features. This iterative method initially uses a simple imputation method such as median imputation, to produce temporary imputation values. In each subsequent iteration it computes for each feature the linear regression model based on all other feature values, and then re-imputes the missing values based on these regression models. The process is terminated upon convergence or after a predetermined number of iterations (Algorithm 1). The authors state that although they do not have a proof of convergence, experiments show fast convergence in most cases (However, in our experiments the method often failed to converge. See Sect. 4.2).

In this paper we present a method similar in spirit to IRMI, formulated as a single optimization problem, and provide an optimization procedure with a guarantee of convergence. This method of Optimized Linear Imputation (OLI) is related in spirit to IRMI in that it performs a linear regression imputation for the missing values of each feature, on the basis of all other features. Our method is defined by a single optimization objective which we then solve using a block coordinate-descent method. Thus our method is guaranteed to converge, which is its most important advantage over IRMI. The OLI method is then compared

---

[1] In this case it would be perhaps more natural to train the model using data pooled over the various copies of the completed data rather than train separate models and average the resulting parameters and structure. This is indeed done artificially in methods such as denoinsing neural nets [37], and has been known to be useful for data imputation [6].

to IRMI as well as other methods using both synthetic, benchmark, and real world datasets.

This paper is an extended version of [29]. The contribution of this paper is as follows: the proposed imputation method is covered in detail, and some of the formulation is revised to promote clarity and simplicity compared to the original conference paper. This extended version also provides a broader review of other linear and non-linear imputation methods.

The rest of the paper is organized as follows: In Sect. 2 we present the novel method of Optimized Linear Imputation (OLI), and a method of optimization which guarantees convergence. We discuss and analyze the relationship to previous methods, and further show that our algorithm may be easily extended to use any form of regularized linear regression.

In Sect. 4 we compare the OLI method to the IRMI, MICE and Median Imputation (MI) methods. Using the same simulation studies as appear in the original IRMI paper, we show that the results of OLI are rather similar to the results of IRMI. With benchmark and real-world datasets we show that our method usually outperforms the alternatives MI and MICE in accuracy, while providing comparable results to IRMI. However, IRMI did not converge in many of these experiments, while our method always provided good results.

Algorithm 1. The IRMI method for imputation of real-valued features (see [34] for more details).

---

input:

- X - data matrix of size $N \times (d+1)$ containing N samples and d features. Zeros in locations of missing values.
- m - missing data mask
- max_iter - maximal number of iterations

output:

- Imputation values

1: $\tilde{X} := median\_impute(X)$ {assigns each missing value the median of its column}
2: **while** not converged and under max_iter iterations **do**
3:    **for** i := 1...d **do**
4:        regression = linear_regression($\tilde{X}_{-i}[!m_i], \tilde{X}_i[!m_i]$)
5:        $\tilde{X}_i[m_i]$ = regression. predict($\tilde{X}_{-i}[m_i]$)
6:    **end for**
7: **end while**
8: **return** $\tilde{X} - X$

---

2   The Optimized Linear Imputation Method

2.1   Notation

We start by listing the notation used throughout the paper.

N	Number of samples
d	Number of features
$x_{i,j}$	The value of the j-th feature in the i-th sample
$m_{i,j}$	Missing value indicators:

$$m_{i,j} = \begin{cases} 1 & x_{i,j} \; is \; missing \\ 0 & otherwise \end{cases}$$

m_i	Indicator vector of missing values for the i-th feature

The following notation is used in the algorithms' pseudo-code:

$A[m]$	The rows of a matrix (or column vector) A where the boolean mask vector m is $True$
$A[!m]$	The rows of a matrix (or column vector) A where the boolean mask vector m is $False$

linear_regression(X, y) A linear regression from the columns of the matrix X to the target vector y, having the following fields:

.parameters: parameters of the fitted model.

.predict(X): the target column y as predicted by the fitted model.

2.2   Optimization Problem

We start by formulating the general problem of data imputation. We will assume that data is given in a matrix X containing missing values in location given by a Boolean mask M (values of the data matrix in missing places can be arbitrary). In the following we state the linear imputation objective as a single optimization problem. First, we construct a design matrix:

$$X = \begin{bmatrix} & & 1 \\ [x_{i,j}(1 - m_{i,j})] & \vdots \\ & & 1 \end{bmatrix} \tag{1}$$

where $x_{i,j}, m_{i,j}$ are the (i, j)-th entry in X and M respectively, and the constant-1 rightmost column is a convenience used later for the intercept terms in the subsequent regression models. Multiplying the data values $x_{i,j}$ by $(1 - m_{i,j})$ simply sets all missing values to zero, keeping non-missing values as they are.

The proposed formulation uses linear regression models as the imputation method, but unlike previous methods does so by means of an optimization problem with a convergence guarantee. The optimization problem approach we present essentially aims to find consistent imputations for all missing values and regression coefficients. By having these values consistent we mean that (a) the imputation values are the values obtained by the regression formulas, and

(b) the regression coefficients are the values that would be computed after the imputations if another iteration of the algorithm was to be applied (i.e. a stationary point of the algorithm). We propose the following optimization formulation:

$$\begin{cases} \min_{A,M} & ||(X+M)A - (X+M)||_F^2 \\ s.t. & m_{i,j} = 0 \Rightarrow M_{i,j} = 0 \\ & M_{i,d+1} = 0 \; \forall i \\ & A_{i,i} = 0 \quad\quad i = 1 \ldots d \\ & A_{i,d+1} = \delta_{i,d+1} \quad \forall i \end{cases} \quad\quad (2)$$

where $||.||_F$ denotes the Frobenius norm.

The objective function defined above is essentially trying to minimize the error of reconstruction of the imputed data $(X+M)$, where each feature (column) is approximated by a linear combination of all other features plus a constant (that is, linear regression of the remaining already-imputed data). The imputation process by which M is defined is guaranteed to leave the non-missing values in X intact, by the first and second constraints which make sure that only missing entries in X have a corresponding non-zero value in M. Therefore:

$$(X+M) = \begin{cases} M & for\; missing\; values \\ X & for\; non\; missing\; values \end{cases}$$

The regression model for each feature is further constrained to use only **other** features, by setting the diagonal values of A to zero (the third constraint). The forth constraint makes sure that the constant-1 rightmost column of the design matrix is copied as-is and therefore does not impact the objective.

We note that all the constraints set variables to constant values, and therefore this can be seen as an unconstrained optimization problem on the remaining set of variables. This set includes the non-diagonal elements of A and the elements of M corresponding to missing values in X. We further note that this is not a convex problem in A, M since it contains the MA factor. In the next section we show a solution to this problem that is guaranteed to converge to a local minimum. This convergence guarantee is the major advantage of the proposed formulation over the prior IRMI method.

2.3   Block Coordinate Descent Solution

We now develop a coordinate descent solution for the proposed optimization problem. Coordinate descent (and more specifically alternating least squares; see for example [2,17,18,33]) algorithms are extremely common in machine learning and statistics, and while don't guarantee convergence to a global optimum (but only to a local optimum), they often preform well in practice.

As stated above, our problem is an unconstrained optimization problem over the following set of variables:

$$\{A_{i,j}|i, j = 1, \ldots, d; i \neq j\} \cup \{M_{i,j}|m_{i,j} = 1\}$$

Keeping this in mind, we re-write the objective function in a form that will facilitate subsequent derivation:

$$L(A, M) = ||(X + M)A - (X + M)||_F^2 \tag{3}$$

$$= \sum_{i=1}^{d} ||(X + M)_{-i}\beta_i - (X + M)_i||_F^2 \tag{4}$$

where C_{-i} denotes the matrix C without its i-th column, C_i the i-th column, and β_i the i-th column of A without the i-th element (recall that the i-th element of the i-th column of A is always zero). The term $(X + M)_{-i}\beta_i$ is therefore a linear combination of all but the i-th column of the matrix $(X + M)$. The sum in (4) is over the first d columns only, since the term added by the rightmost column is zero (see fourth constraint in (2) which enforces the exact copy of the rightmost column).

We now suggest the following coordinate descent algorithm for the minimization of the objective (3) (the method is summarized in Algorithm 2):

1. Fill in missing values using median/mean (or any other) imputation
2. Repeat until convergence:
   (a) Minimize the objective (3) w.r.t. A (compute the columns of the matrix A)
   (b) Minimize the objective (3) w.r.t. M (compute the missing values entries in matrix M).
3. Return M^2

As we will show shortly, step (a) in the iterative part of the proposed algorithm reduces to calculating the linear regression for each feature on the basis of all other features, essentially the same as the first step in the IRMI algorithm [34] Algorithm 1.

Step (b) can be solved either as a system of linear equations or in itself as an iterative procedure, by gradient descent on (3) w.r.t M using (5).

We now begin by briefly showing that step (a) indeed reduces to linear regression. Taking the derivatives of (4) w.r.t the non-diagonal elements of column i of the matrix A we have:

$$\frac{\partial L}{\partial \beta_i} = 2(X + M)_{-i}^T[(X + M)_{-i}\beta_i - (X + M)_i]$$

Setting the partial derivatives to zero gives:

$$(X + M)_{-i}^T[(X + M)_{-i}\beta_i - (X + M)_i] = 0$$
$$\Rightarrow \beta_i = ((X + M)_{-i}^T(X + M)_{-i})^{-1}(X + M)_{-i}^T(X + M)_i$$

---

[2] Alternatively, in order to stay close in spirit to the linear IRMI method, we may prefer to use $(X + M)A$ as the imputed data, meaning the imputed values are in fact derived from the all other features using a linear model. Clearly, at the point of convergence of the algorithm the two are identical.

Algorithm 2. Optimized Linear Imputation (OLI).

---
input:

- X_0 - data matrix of size $N \times d$ containing N samples and d features
- m- missing data mask

output:

- Imputation values

1: $X := median\_impute(X_0)$
2: $M := zeros(N, d)$
3: $A := zeros(d, d)$
4: **while** not converged **do**
5:     **for** $i := 1 \dots d$ **do**
6:         $\beta := linear\_regression(X_{-i}, X_i).parameters$
7:         $A_i := [\beta_1, \dots, \beta_{i-1}, 0, \beta_i, \dots, \beta_d]^T$
8:     **end for**
9:     **while** not converged **do**
10:       $M := M - \alpha[(X + M)A - (X + M)](A - I)^T$
11:       $M[!m] := 0$
12:     **end while**
13:     $X := X + M$
14: **end while**
15: **return** M

---

which is exactly the linear regression coefficients for the *i-th* feature from all other (imputed) features, as claimed.

Next, we obtain the derivatives of the objective function w.r.t M:

$$\nabla_M = \frac{\partial L}{\partial M} = 2[(X + M)A - (X + M)](A - I)^T \tag{5}$$

leading to the following gradient descent algorithm for step (b), the minimization of the objective w.r.t M:

Repeat until convergence:

(i) $M := M - \alpha \nabla_M L(A, M)$
(ii) $\forall_{i,j} : M_{i,j} = M_{i,j} m_{i,j}$

where α is a predefined step size and the gradient is given by (5). Step (ii) above makes sure that only missing values are assigned imputation values[3].

---

[3] Note that this is not a projection step. Recall that the optimization problem is only over elements M_{ij} where x_{ij} is a missing value, encoded by $m_{ij} = 1$. The element-wise multiplication of M by m guarantees that all other elements of M are assigned 0. Effectively, the gradient descent procedure does not treat them as independent variables, as required.

Our proposed algorithm uses a gradient descent procedure for the minimization of the objective (3) w.r.t M. Alternatively, one could use a closed form solution by directly setting the partial derivative to zero. More specifically, let

$$\frac{\partial L}{\partial M} = 0 \tag{6}$$

Substituting (5) into (6), we get

$$M(A - I)(A - I)^T = -X(A - I)(A - I)^T$$

which we rewrite as:

$$MP = Q \tag{7}$$

with the appropriate matrices P, Q. Now, since only elements of M corresponding to missing values of X are optimization variables, only these elements must be set to zero in the derivative (6), and hence only these elements must obey the equality (7). Thus, we have:

$$(MP)_{i,j} = Q_{i,j} \forall i, j | m_{i,j} = 1$$

which is a system of $\sum_{i,j} m_{i,j}$ linear equations in $\sum_{i,j} m_{i,j}$ variables.

2.4 Discussion

In this section we take another look at the IRMI and OLI methods, in order to better understand the difference between them. Specifically, we examine the problem formulations and their implications. We begin by rewriting the IRMI iterative method [34] using the same notation as used for our OLI method. Once on common grounds, we compare the formulations and solutions.

We start by defining an error matrix:

$$E = (X + M)A - (X + M)$$

In the following, E is the error matrix of the linear regression models on the basis of the imputed data. Unlike our method, however, IRMI considers the error only in the non-missing values of the data, leading to the following objective function:

$$L(M, A) = \sum_{i,j | m_{i,j}=0} E_{i,j}^2$$

In order to minimize this loss function, at each step the IRMI method (Algorithm 1) optimizes over a single column of A (which in effect reduces to fitting a single linear regression model), and then assigns as the missing values in the corresponding column of M the values predicted for it by the regression model.

While this heuristic for choosing M is quite effective, it is **not** a gradient descent step and consequently leads to a process with unknown convergence

properties. The main motivation for proposing our method was to fix this undesired property within the same general conceptual framework of linear imputation; namely, propose a method that is similar in spirit, with a convergence
guarantee.

Another advantage of the proposed OLI formulation is the ability to easily
extend it to any regularized linear regression. This can be done by re-writing the
itemized form of the objective (4) as follows:

$$L(A, M) = \sum_i [|||(X + M)_{-i}\beta_i - (X + M)_i||_F^2 + \Omega(\beta_i)]$$

where $\Omega(\beta_i)$ is the regularization term.

Now, assuming that the resulting regression problem can be solved (that is,
minimizing each of the summands in the new objective with a constant M), and
since step (b) of our method remains exactly the same (the derivative w.r.t M
does not change as the extra term does not depend on M), we can use the same
method to solve this problem as well.

Another possible extension is to use kernelized linear regression. In this case
the imputation is preformed in an implicit feature space:

$$L(A, M) = ||\phi((X + M)A) - \phi(X + M)||_F^2$$

This may be useful in cases when the dependencies between the features are
not linear (see a further discussion of this case in Sect. 3).

The method of initialization is another issue deserving further investigation.
Since our procedure converges to a local minimum of the objective, it may be
advantageous to start the procedure from several random initial points, and
choose the best final result. However, since the direct target (missing values) are
obviously unknown, we would need an alternative measure of the "goodness" of
a result. The missing values are usually assumed to be missing at random, so it
would make sense to use the distance between the distributions of known and
imputed values (per feature) as a measure of appropriateness of an imputation.

3   Non-linear Imputation Methods

Linear imputation methods are the family of methods which model a missing
value as a linear combination of other values in the same record (either only
non-missing or both missing and non-missing). Using these methods makes sense
when a notion of a *record* exists in the data, in the sense that a set of measurements refers to the same entity in some way. This is often the case in tabular (or
relational) data, where each row represents information about a specific entity.
In this case, redundancy in the structure of the record often allow linear data
imputation.

However, the structure of the data is often such that non-linear relationships
exist between columns, and thus non-linear methods are required in order to
model and impute missing data.

One of the most utilized non-linear imputation methods uses the method of Expectation Maximization in order to obtain maximum likelihood estimates for the missing data [4]. A major advantage of this approach is the convergence guarantee, and a vast literature regarding statistical properties in various settings and under different assumptions regarding the underlying model.

Deep learning techniques have become overwhelmingly popular in recent years for many machine learning tasks. Indeed, this shift has not skipped the very important task of data imputation.

Stacked Denoising Autoencoders [37] (SDA) is a training method for deep learning models where noise is added to the training examples and fed into the network. The objective is then to recover the original version of the data (prior to having the noise added to it). The main use of the SDA method is for learning representations (see for instance [16,22,23,40]), however in [6] this method is proposed as a means for imputation of traffic data.

In order to use SDA for data imputation, during the training stage a "missing data" mask is randomly selected for each sample, and the corresponding values are zeroed-out or replaced with noise values. The objective function the network is trained with respect to, as in the case of denoising, is the reconstruction error of the output versus the original data. When the trained model is used for imputation, the actual missing data is treated like the mask during training, and the output of the network is used as the imputed data.

The most straightforward version of this method would require a substantial amount of data without any missing elements, since these are used for the training process described above. However, one might use training data which does contain missing elements, and use a loss which takes this into account (essentially by requiring the reconstruction error to be low only for true data).

In image processing, the task of image denoising is to clean up noise in a digital image, which appears either due to noise in the acquisition process (dust, rain, etc.), or as the result of some intentional post-processing (such as overlayed text). Although often treated differently, image denoising is essentially an imputation problem. Here too, denoising autoencoders have been employed successfully [39] to achieve state of the art results.

In the recently proposed learning setting of *Ballpark Learning* [12], an entire column is imputed (or estimated) based on rough group comparisons. In this setting, rather than having column-wise partial information, upper and lower bounds on the proportions of labels in so-called "bags" are used together with constraints on bag differences to obtain an optimization problem yielding the desired imputations of the target column.

Non-linear imputation methods are potentially superior to linear methods, when the linear structure assumption the latter are based on is not a good description of the data. However, when applicable, linear methods have the very desirable advantages of simplicity and interpretability, which arguably is what makes them so popular in practice.

In the next section we present an in-depth evaluation of the OLI method proposed in this paper, and compare it to other linear imputation methods. We start of using synthetic data, them move on to some benchmark datasets.

4 Evaluation

In order to evaluate our method, we compared its performance to other imputation methods using various types of data. We used complete datasets (real or synthetic), and randomly eliminated entries in order to simulate the missing data case. To evaluate the success of each imputation method, we used the mean square error (MSE) of the imputed values as a measure of error. MSE is computed as the mean square distance between stored values (the correct values for the simulated missing values) and the imputed ones.

In Sect. 4.1 we repeat the experimental evaluation from [34] using synthetic data, in order to compare the results of our method to the results of IRMI. In Sect. 4.2 we compare our method to 3 other methods - IRMI, MI (median imputation), and MICE - using standard benchmark datasets from the UCI repository [19]. In Sect. 4.3 we augment the comparisons with an addition new real-life dataset of behavioral modes of migrating storks [30].

For some real datasets in the experiments described below we report that the IRMI method did not converge (and therefore did not return any result). This decision was reached when the MSE of the IRMI method rose at least 6 orders of magnitude throughout the allocated 50 iterations, or (when tested with unlimited iterations) when it rose above the maximum valid number in the system of approximately $1e+308$.

4.1 Synthetic Data

The following simulation studies follow [34] and compare OLI to IRMI. All simulations are repeated 20 times with 10, 000 samples. 5% of all values across records are selected at random and marked as missing. Values are stored for comparison with imputed values. Simulation data is multivariate normal with mean of 1 in all dimensions. Unless stated otherwise, the covariance matrix has 1 in its diagonal entries and 0.7 in the off-diagonal entries.

The aim of the first experiment is to test the relationship between the actual values imputed by the IRMI and OLI methods. The simulation is based on multivariate normal data with 5 dimensions. Results show that the values imputed by the two methods are very near (Fig. 1), with the vast majority of values imputed by the two methods with an absolute difference of up to .02 (compared to the standard deviation of 1.0 in the data. Furthermore, the distribution of imputation error derived from the two methods is identical. Together, these findings point to the similarity in the results these two methods produce.

In the next simulation we test the performance of the two methods as we vary the number of features. The simulation is based on multivariate normal data with 3–20 dimensions. The results (Fig. 2b) show almost identical behavior of

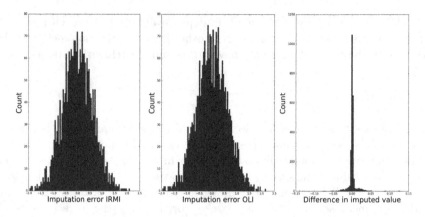

Fig. 1. Distribution of imputation error (imputed − actual) for the IRMI method lLeft). Distribution of imputation error for the OLI method (center). Distribution of the difference in the imputed value between the IRMI and OLI methods (right). Data is 10,000 samples from a 5-dimensional multivariate normal distribution. All columns have a standard deviation of 1.0 and all pairs of columns have a correlation of 0.7, 5.0% of the data was randomly selected and designated as missing.

the IRMI and OLI algorithms, which also coincides with the results presented for IRMI in [34]. Median imputation (MI) is also shown for comparison as baseline. Figure 3 shows a zoom into a small segment of Fig. 2.

As expected, imputing the median (which is also the mean) of each feature for all missing values results in an MSE equal to the standard deviation of the features (i.e., 1). While very close, the IRMI and the OLI methods do not return the exact same imputation values and errors, with an average absolute deviation of 0.053.

Next we test the performance of the two methods as we vary the covariance between the features. The simulation is based on multivariate normal data with 5 dimensions. Non-diagonal elements of the covariance matrix are set to values in the range 0.1–0.9. The results (Fig. 2a) show again almost identical behavior of the IRMI and OLI algorithms. As expected, when the dependency between the feature columns is increased, which is measure by the covariance between the columns (X-axis in Fig. 2a), the performance of the regression-based methods IRMI and OLI is monotonically improving, while the performance of the MI method remain unaltered.

4.2 UCI Datasets

The UCI machine learning repository [19] contains several popular benchmark datasets, some of which have been previously used to compare methods of data imputation [32]. In the current experiment we used the following datasets: *iris* [8], *wine* (white) [3], *Ecoli* [14], *Boston housing* [10], and *power* [35]. Each feature of each dataset was normalized to have mean 0 and standard deviation

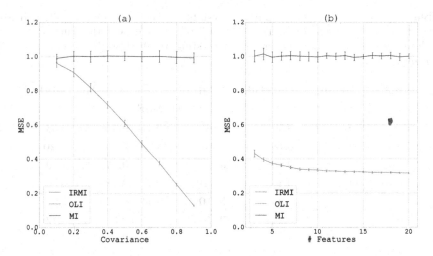

Fig. 2. (a) MSE of the IRMI, OLI and MI methods as a function of the covariance. Data is 5 dimensional multivariate normal. (b) MSE of the IRMI, OLI and MI methods as a function of the dimensionality, with a constant covariance of 0.7 between pairs of features. In both cases error bars represent standard deviation over 20 repetitions.

of 1, in order to make error values comparable between datasets. Categorical features were dropped. For each dataset, 5% of the values were chosen at random and replaced with a missing value indicator. The procedure was repeated 10 times. For these datasets we also consider the MICE method [1] using the *winMice* [15] software.

Overall, the results are quite good, demonstrating the superior ability of the linear methods to impute missing data in these datasets (Table 1, rows 1–5). In the Iris dataset our OLI method achieved an average error identical to IRMI,

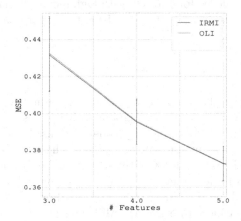

Fig. 3. Zoom into a small part of Fig. 2.

Table 1. Comparison of the imputation results of the IRMI, OLI, MICE and MI methods with 5% missing data. The *converged* column indicates the number of runs in which the IRMI method converged during testing; the MSE of IRMI was calculated for converged repetitions only.

Dataset	# Features	Correlation	IRMI		OLI	MI	MICE
			Converged	MSE			
Iris	4	0.59	9/10	**0.20**	**0.20**	1.00	0.33
Ecoli	7	0.18	9/10	8.26	5.75	1.72	**1.20**
Wine	11	0.18	**0/10**	-	**0.87**	1.05	1.10
Housing	11	0.45	10/10	**0.28**	0.30	1.14	0.56
Power	4	0.45	**3/10**	0.44	**0.47**	1.02	0.88
Storks	20	0.24	**0/10**	-	**0.31**	1.07	0.42

which successfully converged only 9 out of the 10 runs. Both outperformed the MI and MICE standard methods. In the Ecoli dataset both the IRMI and OLI methods performed worse than the alternative methods, with MICE achieving the lowest MSE. In the Wine dataset the IRMI failed to converge in all 10 repetitions, while the OLI method outperformed the MI and MICE methods. The IRMI method outperformed all other methods in the Housing dataset, but failed to converge 7 out of 10 times for the Power dataset.

In summary, in cases where the linear methods were appropriate, with sufficient correlation between the different features (shown in the second column of Table 1), the proposed OLI method was comparable to the IRMI method with regard to mean square error of the imputed values when the latter converged, and superior in that it always converges and therefore always returns a result.

While the IRMI method achieved slightly better results than OLI in some cases, its failure to converge in others gives the OLI method the edge. Overall, better results were achieved for datasets with high mean correlation between features, as expected when using methods utilizing the linear relationships between features.

4.3   Storks Behavioral Modes Dataset

In the field of Movement Ecology, readings from accelerometers placed on migrating birds are used for both supervised [26] and unsupervised [27,28] learning of behavioral modes. In the following experiment we used a dataset of features extracted from 3815 such measurements. As with the UCI datasets, 10 repetitions were performed, each with 5% of the values randomly selected and marked as missing. Results (Table 1, final row) of this experiment highlight the relative advantage of the OLI method. While the IRMI method failed to converge in all 10 repetitions, OLI achieved an average MSE considerably lower than the MI baseline, and also outperformed the MICE method.

5   Conclusion

Since the problem of missing values often haunts real-word datasets, while most data analysis methods are not designed to deal with this problem, imputation is a necessary pre-processing step whenever discarding entire records is not a viable option. Here we proposed an optimization-based linear imputation method that augments the IRMI [34] method with the property of guaranteed convergence, while staying close in spirit to the original method. Since our method converges to a local optimum of a different objective function, the two methods should not be expected to converge to the same value exactly. However, simulation results show that the results of the proposed method are generally similar (nearly identical) to IRMI when the latter does indeed converge.

The contribution of our paper is two-fold. First, we suggest an optimization problem based method for linear imputation and an algorithm that is guaranteed to converge. Second, we show how this method can be extended to use any number of methods of regularized linear regression. Unlike matrix completion methods [38], we do not have a low rank assumption. Thus, OLI should be preferred when data is expected to have some linear relationships between features and when IRMI fails to converge, or alternatively, when a guarantee of convergence is important (for instance in automated processes).

References

1. Buuren, S., Groothuis-Oudshoorn, K.: MICE: multivariate imputation by chained equations in R. J. Stat. Softw. **45**(3) (2011)
2. Comon, P., Luciani, X., De Almeida, A.L.: Tensor decompositions, alternating least squares and other tales. J. Chemometr. **23**(7–8), 393–405 (2009)
3. Cortez, P., Cerdeira, A., Almeida, F., Matos, T., Reis, J.: Modeling wine preferences by data mining from physicochemical properties. Decis. Support Syst. **47**(4), 547–553 (2009)
4. Dempster, A.P., Laird, N.M., Rubin, D.B.: Maximum likelihood from incomplete data via the EM algorithm. J. Royal Stat. Soc. Ser. B (Methodol.) **39**(1), 1–38 (1977)
5. Donders, A.R.T., van der Heijden, G.J., Stijnen, T., Moons, K.G.: Review: a gentle introduction to imputation of missing values. J. Clin. Epidemiol. **59**(10), 1087–1091 (2006)
6. Duan, Y., Yisheng, L., Kang, W., Zhao, Y.: A deep learning based approach for traffic data imputation. In: 2014 IEEE 17th International Conference on Intelligent Transportation Systems (ITSC), pp. 912–917. IEEE (2014)
7. Engels, J.M., Diehr, P.: Imputation of missing longitudinal data: a comparison of methods. J. Clin. Epidemiol. **56**(10), 968–976 (2003)
8. Fisher, R.A.: The use of multiple measurements in taxonomic problems. Ann. Eugen. **7**(2), 179–188 (1936)
9. García-Laencina, P.J., Sancho-Gómez, J.L., Figueiras-Vidal, A.R.: Pattern classification with missing data: a review. Neural Comput. Appl. **19**(2), 263–282 (2010)
10. Harrison, D., Rubinfeld, D.L.: Hedonic housing prices and the demand for clean air. J. Environ. Econ. Manag. **5**(1), 81–102 (1978)

11. Heitjan, D.F., Basu, S.: Distinguishing missing at random and missing completely at random. Am. Stat. **50**(3), 207–213 (1996)

12. Hope, T., Shahaf, D.: Ballpark learning: estimating labels from rough group comparisons. In: Frasconi, P., Landwehr, N., Manco, G., Vreeken, J. (eds.) ECML PKDD 2016. LNCS (LNAI), vol. 9852, pp. 299–314. Springer, Cham (2016). https://doi.org/10.1007/978-3-319-46227-1_19

13. Horton, N.J., Kleinman, K.P.: Much ado about nothing: a comparison of missing data methods and software to fit incomplete data regression models. Am. Stat. **61**(1), 79–90 (2007)

14. Horton, P., Nakai, K.: A probabilistic classification system for predicting the cellular localization sites of proteins. In: Ismb. vol. 4, pp. 109–115 (1996)

15. Jacobusse, G.: WinMICE users manual. TNO quality of life, Leiden (2005). http://www.multiple-imputation.com

16. Kandaswamy, C., Silva, L.M., Alexandre, L.A., Sousa, R., Santos, J.M., de Sá, J.M.: Improving transfer learning accuracy by reusing stacked denoising autoencoders. In: 2014 IEEE International Conference on Systems, Man and Cybernetics (SMC), pp. 1380–1387. IEEE (2014)

17. Kim, H., Park, H.: Nonnegative matrix factorization based on alternating nonnegativity constrained least squares and active set method. SIAM J. Matrix Anal. Appl. **30**(2), 713–730 (2008)

18. Kroonenberg, P.M., De Leeuw, J.: Principal component analysis of three-mode data by means of alternating least squares algorithms. Psychometrika **45**(1), 69–97 (1980)

19. Lichman, M.: UCI machine learning repository (2013). http://archive.ics.uci.edu/ml

20. Little, R.J.: A test of missing completely at random for multivariate data with missing values. J. Am. Stat. Assoc. **83**(404), 1198–1202 (1988)

21. Little, R.J., Rubin, D.B.: Statistical Analysis with Missing Data. Wiley, Hoboken (2014)

22. Lu, X., Tsao, Y., Matsuda, S., Hori, C.: Speech enhancement based on deep denoising autoencoder. In: Interspeech, pp. 436–440 (2013)

23. Masci, J., Meier, U., Cireşan, D., Schmidhuber, J.: Stacked convolutional autoencoders for hierarchical feature extraction. In: Honkela, T., Duch, W., Girolami, M., Kaski, S. (eds.) ICANN 2011. LNCS, vol. 6791, pp. 52–59. Springer, Heidelberg (2011). https://doi.org/10.1007/978-3-642-21735-7_7

24. Pigott, T.D.: A review of methods for missing data. Educ. Res. Eval. **7**(4), 353–383 (2001)

25. Raghunathan, T.E., Lepkowski, J.M., Van Hoewyk, J., Solenberger, P.: A multivariate technique for multiply imputing missing values using a sequence of regression models. Surv. Methodol. **27**(1), 85–96 (2001)

26. Resheff, Y.S., Rotics, S., Harel, R., Spiegel, O., Nathan, R.: Accelerater: a web application for supervised learning of behavioral modes from acceleration measurements. Mov. Ecol. **2**(1), 25 (2014)

27. Resheff, Y.S., Rotics, S., Nathan, R., Weinshall, D.: Matrix factorization approach to behavioral mode analysis from acceleration data. In: IEEE International Conference on Data Science and Advanced Analytics (DSAA), 36678 2015, pp. 1–6. IEEE (2015)

28. Resheff, Y.S., Rotics, S., Nathan, R., Weinshall, D.: Topic modeling of behavioral modes using sensor data. Int. J. Data Sci. Anal. **1**(1), 51–60 (2016)

29. Resheff, Y.S., Weinshal, D.: Optimized linear imputation. In: Proceedings of the 6th International Conference on Pattern Recognition Applications and Methods, ICPRAM, vol. 1, pp. 17–25 (2017)

30. Rotics, S., Kaatz, M., Resheff, Y.S., Turjeman, S.F., Zurell, D., Sapir, N., Eggers, U., Flack, A., Fiedler, W., Jeltsch, F., et al.: The challenges of the first migration: movement and behaviour of juvenile vs. adult white storks with insights regarding juvenile mortality. J. Anim. Ecol. **85**(4), 938–947 (2016)

31. Rubin, D.B.: Multiple imputation after 18+ years. J. Am. Stat. Assoc. **91**(434), 473–489 (1996)

32. Schmitt, P., Mandel, J., Guedj, M.: A comparison of six methods for missing data imputation. J. Biom. Biostat. (2015)

33. Takane, Y., Young, F.W., De Leeuw, J.: Nonmetric individual differences multidimensional scaling: an alternating least squares method with optimal scaling features. Psychometrika **42**(1), 7–67 (1977)

34. Templ, M., Kowarik, A., Filzmoser, P.: Iterative stepwise regression imputation using standard and robust methods. Comput. Stat. Data Anal. **55**(10), 2793–2806 (2011)

35. Tüfekci, P.: Prediction of full load electrical power output of a base load operated combined cycle power plant using machine learning methods. Int. J. Electr. Power Energy Syst. **60**, 126–140 (2014)

36. Van Buuren, S., Oudshoorn, K.: Flexible multivariate imputation by MICE. TNO Prevention Center, Leiden, The Netherlands (1999)

37. Vincent, P., Larochelle, H., Lajoie, I., Bengio, Y., Manzagol, P.A.: Stacked denoising autoencoders: learning useful representations in a deep network with a local denoising criterion. J. Mach. Learn. Res. **11**, 3371–3408 (2010)

38. Wagner, A., Zuk, O.: Low-rank matrix recovery from row-and-column affine measurements. arXiv preprint arXiv:1505.06292 (2015)

39. Xie, J., Xu, L., Chen, E.: Image denoising and inpainting with deep neural networks. In: Advances in Neural Information Processing Systems, pp. 341–349 (2012)

40. Zhou, G., Sohn, K., Lee, H.: Online incremental feature learning with denoising autoencoders, Ann Arbor (2012)

Condensing Deep Fisher Vectors: To Choose or to Compress?

Sarah Ahmed[(⊠)] and Tayyaba Azim[(⊠)]

Center of Excellence in IT, Institute of Management Sciences, Peshawar, Pakistan
ssarahahmedd@gmail.com, tayyaba.azim@imsciences.edu.pk

Abstract. *Feature selection* and *dimensionality reduction* are the two popular off-the-shelf techniques in practice for reducing data's high dimensional memory footprint and thus making it amenable for large scale visual retrieval and classification. In this paper, we show that *feature compression* is a better choice than *feature selection* when dealing with large scale retrieval of high dimensional Fisher vectors derived from deep or shallow stochastic models such as restricted Boltzmann machine (RBM). The dimensionality of the Fisher vectors is proportional to the size of the architecture from which they are drawn. As the number of hidden units in RBM increases, the dimensionality of the Fisher vectors also scales accordingly, thus increasing storage requirements as well as causing overfitting during classification. In order to tackle these challenges, we compare the performance of feature compression and feature selection techniques and suggest the use of compression methods on available Fisher encodings. We have based our diagnostics on *multi-collinearity* evaluation metrics and justify the use of the proposed feature condensation method using feature visualisations and classification accuracy on benchmark data set.

1 Introduction

Large scale image classification and retrieval has received an increasing attention over the last decade due to the availability of large amount of multimedia data on the web and the growing need to mine information of interest from these large image repositories. Where on one end, we have witnessed improvements in the hardware to efficiently store and process such massively growing data sets, efforts have also been made at the algorithmic level to come up with speedy retrieval techniques that are human competitive in perception and image understanding tasks. These algorithms rely specifically on how the images are represented semantically in a feature space that makes them discriminant as well as retrievable for later use. In this regard, one of the most popular approaches to represent images through mid level features is *bag of visual words (BoW)* approach [1] that converts the visual vocabulary built in *low level* feature space into *intermediate* representations of fixed size. These features have conventionally been used to train a non-linear classifier like support vector machines (SVM) and have consistently shown to outperform other methods in successive image

© Springer International Publishing AG, part of Springer Nature 2018
M. De Marsico et al. (Eds.): ICPRAM 2017, LNCS 10857, pp. 80–98, 2018.
https://doi.org/10.1007/978-3-319-93647-5_5

classification evaluations like PASCAL VOC [2] and CALTECH 101/256 [3,4]. Despite its success, an important limitation of this approach lies in its inability to scale to large amounts of training data. Merely computing the kernel matrix in non-linear SVM requires $O(n^2 d)$ calculations, where n is the number of training examples and d represents the dimensionality of image representation. In BoW model with a large codebook and spatial pyramid, the value of d can be as large as hundreds of thousands. When n also becomes large, non-linear SVM becomes computationally intractable. The cost of non-linear SVM in the test phase is also very high, i.e. $O(dn_{sv})$, where n_{sv} denotes the total number of support vectors which grows as the data set becomes large. All these factors make non-linear SVM unattractive for large scale image classification and retrieval problems. To achieve both high computational efficiency and competitive performance, recent research has shifted its focus on either replacing the classifier or choosing a better encoding scheme [5–9] that can perform well even with a linear classifier.

The *Fisher kernel* (FK) framework introduced by Jaakola and Haussler [10] and applied by Perronin and Dance [11] to image classification task is an extension of the initial bag of words (BoW) idea explained in detail ahead (Sect. 2). The FK combines the benefits of generative and discriminative approaches to pattern classification by deriving a kernel from a generative probability model of the data. The Fisher features have shown to overcome the limitations of BoW approach [12] and have yielded competitive results for large scale image classification and retrieval tasks [13,14]. Another prominent feature of the Fisher vector is that it performs very well even with a simple linear classifier using techniques such as stochastic gradient descent method. However, these recommended Fisher features have high dimensionality and in combination with a large number of examples could pose serious computational and storage constraints [15]. This problem has been tackled by either using standard compression techniques [15] or through feature selection methods [16] that reduce the signature length of each image to acquire less storage and quick retrieval results.

This paper is an extended version of our previous work [17] that takes into account efficient ways of condensing Fisher vectors derived from restricted Boltzmann machine (RBM) [18] with minimal loss in classification performance. The dimensionality of the Fisher vectors derived from deep models has an intrinsic relationship with the number of hidden units of the model. As the number of hidden units increases, the length of the encoded Fisher vectors also scales. See Table 1 to understand the growth rate of deep Fisher vectors. It has already been shown in literature that when Fisher kernel is derived from a large restricted Boltzmann machine with thousands of units, the classifier suffers from overfitting [18,19]. If we are to take advantage of large generative models for learning efficient classifiers, some feature condensation mechanism must be utilised to make the approach practical for large scale image retrieval and classification. *This paper focuses on techniques to reduce the Fisher feature dimensionality and hence the storage cost and computational overhead required for large scale visual retrieval problems.* Compared to our previous work [17], where only compression techniques were explored, this paper also investigates the use of feature

selection schemes on deep Fisher vectors. The contributions unique to this work
are as follows: (1) Comparison of compression techniques with feature selection
methods to explore their suitability for large scale retrieval problems. The feature compression techniques used are: Parametric t-SNE, autoencoder, principal
component analysis (PCA) and spectral hashing (SH). For feature selection,
the following filters and wrappers are used: Maximum relevance and minimum
redundancy (MRMR), mutual information, SVM recursive feature elimination
(RFE) and random forest, (2) Assessment of collinearity in deep Fisher vectors
via two diagnostic tests: (a) Variance inflation factor (VIF) and (b) Conditional
indices, (3) Empirical analysis of state of the art classifiers using the above mentioned feature compression and selection techniques, (4) Analysis of the effect of
normalisation schemes on feature compression and feature selection methods.

Table 1. Growth of Fisher vector's length in MNIST data set where the images have
dimensionality 28 × 28.

No of hidden units in RBM	1	5	10	100	1000
Fisher vector's length ($l = |\mathbf{v} \times \mathbf{h}| + |\mathbf{v}| + |\mathbf{h}|$)	1569	4709	8634	79284	785784

2   The Fisher Kernel Framework

The Fisher kernel framework [10] proposes to use the power of generative models
$P(\mathbf{x}|\boldsymbol{\theta})$ in kernel methods by computing Fisher scores using gradients of the log
likelihood of the data, $\mathbf{x}$ with respect to the model parameters, $\boldsymbol{\theta}$. The derived
kernel function uses these Fisher scores/vectors in the following form:

$$K(\mathbf{x}_i, \mathbf{x}_j) = \phi_{\mathbf{x}_i}^T \mathbf{J}^{-1} \phi_{\mathbf{x}_j}, \tag{1}$$

where $\mathbf{J}$ is the covariance matrix of the Fisher scores, $\phi_{\mathbf{x}}$ and is regarded as the
Fisher Information matrix, i.e.

$$\mathbf{J} = \mathrm{E}\left[\phi_{\mathbf{x}}^T \phi_{\mathbf{x}}\right]_{P(\mathbf{x})}, \text{ where } \phi_{\mathbf{x}} = \nabla_{\boldsymbol{\theta}}\left[\log P(\mathbf{x}; \boldsymbol{\theta})\right]. \tag{2}$$

Fisher kernel works on the intuition that two similar structured objects
should have similar gradients in the parameter space of the generative model.
The computation of Fisher information matrix is generally considered immate-
rial [10] and is often ignored in practice by replacing it with an identity matrix, I.
However, some of the literature on the classification systems has also shown good
discrimination results by using approximations of the information matrix in kernel computation [20]. Examples of such approximations include restricted forms
of covariance matrix, such as a diagonal covariance matrix ($\mathbf{J} = \mathrm{diagonal}(\sigma^2)$)
or isotropic Gaussians ($\mathbf{J} = \sigma^2 I$). Fisher kernel, once derived from a generative
probability model, $P(\mathbf{x}|\boldsymbol{\theta})$ is capable of being embedded into any discriminative

classifier such as support vector machines (SVM), linear discriminant analysis (LDA), neural networks, etc.

In this work, we have taken a restricted Boltzmann machine (RBM) [21] to derive Fisher scores. A restricted Boltzmann machine is a bipartite graph in which the visible units that represent observations are connected to binary stochastic hidden units using undirected weight connections. The hidden units are used to discover useful features or patterns from the data fed to the visible layer during training. The probability of a joint configuration over both visible and hidden units depends on the energy of that joint configuration compared with the energy of all other joint configurations:

$$P(\mathbf{v}, \mathbf{h}; \boldsymbol{\theta}) = \frac{1}{Z(\theta)} \exp(E(\mathbf{v}, \mathbf{h}, \boldsymbol{\theta})), \tag{3}$$

$$Z(\boldsymbol{\theta}) = \sum_{\mathbf{v}, \mathbf{h}} \exp(E(\mathbf{v}, \mathbf{h}, \boldsymbol{\theta})).$$

The parameters of this energy based model are learnt by performing stochastic gradient descent learning on the empirical negative log-likelihood of the training data. A guide to initialise and optimise these parameters, $\boldsymbol{\theta} = \{W, \mathbf{a}, \mathbf{b}\}$ is given by Hinton [22]. The Fisher scores $\phi_{\mathbf{x}}$ derived from the gradients of the log likelihood of the data $\mathbf{x}$ with respect to RBM parameters $\boldsymbol{\theta} = \{W, \mathbf{a}, \mathbf{b}\}$ are given as below:

$$\nabla_{\theta} \log P(\mathbf{x_n}|\boldsymbol{\theta}) = \left[S_{[\mathbf{n}]} \mid Q_{[\mathbf{n}]} \mid Z_{[\mathbf{n}]} \right], \text{ where} \tag{4}$$

$$S_{[\mathbf{n}]} = \nabla_W \log P(\mathbf{x_n}|\boldsymbol{\theta}) = \langle \mathbf{vh}^T \rangle_{P_{data}} - \langle \mathbf{vh}^T \rangle_{P_{model}}, \tag{5}$$

$$Q_{[\mathbf{n}]} = \nabla_{\mathbf{a}} \log P(\mathbf{x_n}|\boldsymbol{\theta}) = \langle \mathbf{h} \rangle_{P_{data}} - \langle \mathbf{h} \rangle_{P_{model}}, \tag{6}$$

$$Z_{[\mathbf{n}]} = \nabla_{\mathbf{b}} \log P(\mathbf{x_n}|\boldsymbol{\theta}) = \langle \mathbf{v} \rangle_{P_{data}} - \langle \mathbf{v} \rangle_{P_{model}}. \tag{7}$$

3   The Fisher Vector Normalisation

In this section, we describe the normalisation scheme required for achieving competitive classification performance with a discriminative classifier using deep Fisher vectors. Feature normalisation is a preprocessing step required to re scale the features in a fixed range. We have applied Min-Max normalisation technique [23] on the Fisher vectors that transform the derived Fisher scores in the range [0, 1]. If $\mathbf{x}$ is an n-dimensional feature vector, the Min-Max normalisation is computed by using the following linear interpretation formula:

$$\mathbf{x}_{norm} = (\mathbf{x}_i - \mathbf{x}_{min})/(\mathbf{x}_{max} - \mathbf{x}_{min}), \tag{8}$$

where $\mathbf{x}_{min}$ and $\mathbf{x}_{max}$ represent the minimum and maximum values across all dimensions for each image vector, $\mathbf{x}$ respectively. The normalised Fisher vector, $\mathbf{x}_{norm}$ has the same dimensionality as that of the Fisher vector, $\mathbf{x}$. The Min-Max normalisation has the advantage of preserving relationship between the original

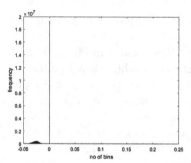

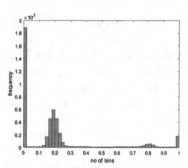

(a) Before normalisation, FVs derived from RBM with 1 hidden unit on MNIST data set.

(b) After normalisation, FVs derived from RBM with 1 hidden unit on MNIST data set.

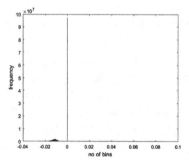

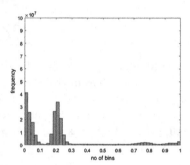

(c) Before normalisation, FVs derived from RBM with 5 hidden units on MNIST data set.

(d) After normalisation, FVs derived from RBM with 5 hidden units on MNIST data set.

Fig. 1. Histogram of Fisher vector features derived from RBM before and after the application of Min-Max normalization [17].

data values and is capable of suppressing the effect of outliers by bounding the range of the data that yields smaller standard deviations. The effect of Min-Max normalisation on deep Fisher vectors could be observed in Fig. 1. Conventionally, the recommended Fisher vectors for large scale retrieval are derived from Gaussian mixture model (GMM) and deploy L2-normalisation scheme to improve their classification performance. We checked the L2 and L1 normalisation techniques for scaling *deep Fisher vectors* but could not find any improvement in the discriminative performance as realised through Min-Max normalisation.

4   Compression Techniques

In this section, we explore the use of following off-the shelf compression techniques for reducing the dimensions of normalised Fisher vectors: Principal component analysis (PCA), spectral Hashing (SH), autoencoder and parametric t-SNE.

4.1   Principal Component Analysis (PCA)

Principal components analysis (PCA) is a linear dimensionality reduction technique that brings out strong patterns in the data set by emphasising disparity in its features through transforming correlated variables into un-correlated variables, also known as *principal components*. The technique embeds the data into a linear subspace M of lower dimensionality describing as much of the variance in the data set as possible. This goal is achieved by computing the covariance matrix $cov(X)$ of data-set X from which the *eigenvectors* and *eigen values* are computed.

Mathematically, PCA selects the linear mapping M that maximises the following cost function: $trace(M^T cov(X)M)$. It can be shown that this linear mapping M is formed by d principal eigen-vectors of the sample covariance matrix of the standardised (zero-mean) data to solve the eigen problem.

$$cov(X)M = \lambda M. \tag{9}$$

The eigen problem is solved for the d principal eigenvalues, λ. The eigen vector corresponding to the largest eigen value gives the direction of greatest variation, similarly the eigen vector with second largest eigen value corresponds to the direction of second highest variation and so on. The low-dimensional data representations Y of the data points X are computed by mapping them onto the linear basis M, i.e. $Y = XM$.

4.2   Spectral Hashing (SH)

Spectral hashing (SH) algorithm [24] is a non-linear dimensionality reduction technique that uses Gaussian kernel to find a binary encoding that minimises the Hamming distance between similar pairs of binary codes. The method works on the intuition that points far apart in the original Euclidean space are also far apart in the Hamming space and vice-versa. The solution for spectral hashing method is simply based on a subset of thresholded eigen vectors of the Laplacian of similarity graph [24].

4.3   Autoencoder

Autoencoder [25] uses a multi-layer stochastic network to transform high-dimensional data into a low-dimensional code and a similar decoder network to recover the original data from the compressed code. The algorithm starts with random weights in the two networks (encoder and decoder) and then trains the two together by minimising the discrepancy between the original data and its reconstruction. The required gradients are easily obtained by using the chain rule to back-propagate error derivatives first through the decoder network and then through the encoder network. Autoencoder is a non-linear generalisation of PCA which can be modelled using a two layer network called restricted Boltzmann machine in which stochastic, binary pixels are connected to stochastic, binary feature detectors using symmetrically weighted connections. The pixels correspond to visible units of the RBM because their states are observed; the feature detectors correspond to hidden units. A joint configuration $(\mathbf{v}, \mathbf{h})$ of the visible and hidden units have the energy:

$$E(\mathbf{v}; \mathbf{h}) = -\sum_{i \in \text{pixels}} b_i v_i - \sum_{j \in features} b_j h_j - \sum_{i,j} v_i h_j w_{i,j}. \tag{10}$$

4.4   Parametric t-SNE

Parametric t-distributed stochastic neighbor embedding (t-SNE) [26] is an unsupervised dimensionality reduction technique which learns a parametric mapping between the high-dimensional and low-dimensional spaces such that the local structure of the data is preserved. In parametric t-SNE, the mapping $f : X \to Y$ from the data space X to the low-dimensional latent space Y is parametrised by means of a feed-forward neural network with weights W. The training procedure is inspired by an autoencoder based on restricted Boltzmann machine (RBM) that operates in three main stages: (1) First, a stack of RBMs is trained, (2) Next, the stack of RBMs is used to construct a pre-trained neural network, and (3) At the end, the pre-trained network is fine-tuned using back-propagation to minimise the cost function that retains local structure of the data in latent space by minimising the Kullback-Leibler (KL) divergence between the probabilities signifying pairwise distances between examples.

5   Feature Selection Techniques

In this section, we briefly discuss the feature selection methods applied to select small subset of features from the original deep Fisher vectors for achieving better classification accuracy.

5.1   Conditional Mutual Information (MI)

Conditional mutual information [27] measures dependency between two features and picks those features which maximise their mutual information with the class to predict. The mutual information of two variables X and Y can be defined as:

$$I(X; Y) = \sum_{y \in Y} \sum_{x \in X} p(x, y) \log \left(\frac{p(x, y)}{p(x)p(y)} \right). \tag{11}$$

Here $p(x, y)$ is the joint probability distribution of function X and Y while $p(x)$ and $p(y)$ are the marginal probability density functions of X and Y. Mutual information measures the sharing information of both X and Y in order to reduce uncertainity in one variable by observing the other variable. If X and Y are dependent variables, then mutual information is greater than 0 and if both X and Y are independent then MI will be 0. In conditional mutual information, score table is updated as features are selected based on their conditional mutual information.

5.2   Minimum Redundancy and Maximum Relevance (MRMR)

MRMR is a filter based selection technique [28] that ranks features by maximising mutual information between the joint probability distribution of the selected features and the classification variable after calculating the minimum redundancy and maximum relevancy of features. The relevancy of feature set S for the class c can be calculated as:

$$D(S,C) = \frac{1}{|S|} \sum_{f_i \in s} I(f_i; c). \tag{12}$$

The redundancy of all features in the set S can be calculated as:

$$R(S) = \frac{1}{|S|^2} \sum_{f_i, f_j \in s} I(f_i; f_j). \tag{13}$$

The feature selection criterion of MRMR involves combination of both these measures that maximise relevance and minimise redundancy of features as follows:

$$MRMR = \max_s \left[\frac{1}{|S|} \sum_{f_i \in s} I(f_i; c) - \frac{1}{|S|^2} \sum_{f_i, f_j \in s} I(f_i; f_j) \right]. \tag{14}$$

5.3   SVM-Recursive Feature Elimination (SVM-RFE)

SVM-RFE is an embedded pruning method [29] that trains a model with all features and removes insignificant features by setting the coefficients associated with these features to 0. It is an iterative process of the backward removal of features carried out by computing the weights of all features and sorting them according to their weights. The technique consist of three main stages: (1) In the first stage, the classifier is trained on the data-set with all the features, (2) In the second stage, the features are given weights and are sorted accordingly setting up their rank. The weight assigning procedure is repeated iteratively and the list of features is assembled according to the order of the weights, (3) In the last stage, the features with the smallest weights are eliminated in-order to retain significant impact of the feature variables.

5.4   Random Forest (RF)

Random forest deploys a combination of decision tree classifiers such that the performance of each tree depends on the values of a random selection of features used to split each node; the random feature vector is sampled independently and with the same probability distribution for all trees in the forest [30]. The random selection of features used for splitting each node yields error rates which are monitored to measure strength and correlation between features. The accuracy of a random forest depends on the strength of the individual tree classifiers and a measure of the dependence between them. According to the central limit theorem, the generalisation error of the forests converges to a limit as the number of trees in the forest becomes very large. The randomness used in tree construction aims for low correlation ρ while maintaining reasonable strength.

6 Multi-collinearity Assessment Diagnostics

This section discusses multi-collinearity evaluation metrics used to assess the possibility of inter-correlation or inter-association among the features of Fisher vectors derived from RBM.

6.1 Variance Inflation Factor (VIF)

Variance inflation factor shows how much is the variance (standard error) inflated due to the existence of correlation between independent variables in a model. Mathematically, it is defined as the reciprocal of *tolerance* given as $1 - R_i^2$, where R_i corresponds to the value predicted by regressing $i - th$ variable by the rest of independent variables. A tolerance close to 1 means that there is little multi-collinearity, whereas a value close to 0 suggests that multi- collinearity may be a threat. Conversely, the variance inflation factor is elaborated by the following piecewise function:

$$VIF \begin{cases} \approx 0 & \implies \text{Moderate to null multi-collinearity} \\ >5 & \implies \text{High multi-collinearity} \end{cases}$$

The threshold for large VIF values is taken at 5 in this paper, however some of the literature also uses 10 as a threshold, i.e. 0.10 tolerance factor to indicate multi-collinearity among independent variables.

6.2 Condition Indices

Condition index calculates the collinearity of combination of variables in the data set by calculating relative size of the eigen values of data matrix X [31]. In order to calculate eigen values, the method applies singular value decomposition (SVD) of the $n \times p$ data matrix X and computes condition indices as below:

$$CI_i = \sqrt{\frac{\lambda_{max}}{\lambda_i}} \tag{15}$$

where λ denotes the eigen values of the correlation matrix signifying variance of the linear combination of independent variables. Condition indices between 30 and 100 indicate moderate to strong collinearity between the features.

7   Experiments

In order to evaluate the classification performance of condensed Fisher vectors (FV), we applied four different compression and feature selection schemes dis-

Table 2. Accuracy of k-NN, random forest and SVM classifiers on un-normalised Fisher vectors derived from RBM with 1 hidden unit using MNIST data set.

Classifiers	k-NN				Random Forest				SVM (Linear Kernel)			
Feature Selection Techniques	2D	10D	20D	Full D	2D	10D	20D	Full D	2D	10D	20D	Full D
MRMR	24%	62%	80%		30%	73%	86%		30%	61%	75%	
MI	23%	46%	65%		27%	59%	76%		27%	44%	60%	
Random Forest	16%	43 %	63%	96%	21%	58%	73%	96%	19%	39 %	59%	90%
SVM-RFE	20%	40%	58%		25%	53%	70%		23%	39%	51%	
Classifiers	k-NN				Random Forest				SVM (Linear Kernel)			
Compression Techniques	2D	10D	20D	Full D	2D	10D	20D	Full D	2D	10D	20D	Full D
Parametric t-SNE	10%	10%	32%		12%	12%	53%		11%	11%	11%	
Autoencoder	9.8%	9.8%	9.8%		10%	10%	10%		11%	11%	11%	
PCA	9.9%	18%	14%	96%	10%	17%	12%	96%	10%	19 %	13%	90%
Spectral Hashing (Nbits= 16, 80, 160 for 2D, 10D and 20D respectively.)	19%	9.9%	10%		20%	11%	12%		13%	10%	11%	

Table 3. Accuracy of k-NN, random forest and SVM classifiers on normalised Fisher vectors derived from RBM with 1 hidden unit using MNIST data set.

Classifiers	k-NN				Random Forest				SVM (Linear Kernel)			
Feature Selection Techniques	2D	10D	20D	Full D	2D	10D	20D	Full D	2D	10D	20D	Full D
MRMR	19%	57%	79%		25%	68%	85%		27%	59%	75%	
MI	19%	45%	65%		24%	58%	74%		23%	47%	63%	
Random Forest	16%	40 %	60 %	96%	21 %	54%	71%	96%	20%	43%	60%	90%
SVM-RFE	19%	38%	57%		24%	53%	69%		20%	40%	51%	
Classifiers	k-NN				Random Forest				SVM (Linear Kernel)			
Compression Techniques	2D	10D	20D	Full D	2D	10D	20D	Full D	2D	10D	20D	Full D
Parametric t-SNE	86%	94%	94.2%		89.6%	95.5%	95.7%		76%	87%	92%	
Autoencoder	78%	89%	96%		86%	95%	95%		65%	92%	89%	
PCA	27%	67%	68%	96%	31%	68%	67%	96%	38%	69 %	64%	90%
Spectral Hashing (Nbits= 16, 80, 160 for 2D, 10D and 20D respectively.)	66%	50%	44%		26%	40%	34%		16%	30%	32%	

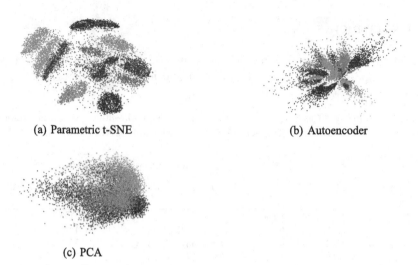

(a) Parametric t-SNE   (b) Autoencoder

(c) PCA

Fig. 2. Visualisation of normalised compressed Fisher scores derived from RBM with 1 hidden unit [17].

Table 4. Accuracy of k-NN, random forest and RBF kernel SVM classifier on unnormalised Fisher vectors derived from RBM with 5 hidden units using MNIST data set.

Classifiers	k-NN				Random Forest				SVM (Linear Kernel)			
Feature Selection Techniques	2D	10D	20D	Full D	2D	10D	20D	Full D	2D	10D	20D	Full D
MRMR	23%	61%	80%		30%	73%	86%		30%	59%	71%	
MI	20%	28%	32%		22%	36%	42%		23%	28%	35%	
Random Forest	18%	39 %	51 %	96%	25 %	53 %	64%	96%	25%	40%	48%	90%
SVM-RFE	18%	31%	34%		23%	42%	47%		21%	33%	38%	

Classifiers	k-NN				Random Forest				SVM (Linear Kernel)			
Compression Techniques	2D	10D	20D	Full D	2D	10D	20D	Full D	2D	10D	20D	Full D
Parametric t-SNE	10%	10%	10%		12%	12 %	12%		11%	11%	11%	
Autoencoder	9.8%	9.8%	9.8%		12%	12%	12%		11%	11%	11%	
PCA	10%	19 %	14%	96%	10%	17%	12%	96%	10%	20 %	15%	90%
Spectral Hashing (Nbits= 16, 80, 160 for 2D, 10D and 20D respectively.)	10%	8%	10%		11%	9%	10%		12%	10%	11%	

cussed in Sects. 4 and 5 and calculated their classification accuracy with the following standard classifiers: k-nearest neighbour, random forest and support vector machines (SVM). The benchmark data set used for performing experiments is MNIST [32]. The MNIST data set consists of 28×28 dimensional gray scale images with 60,000 digits in the training and 10,000 digits in the test sets.

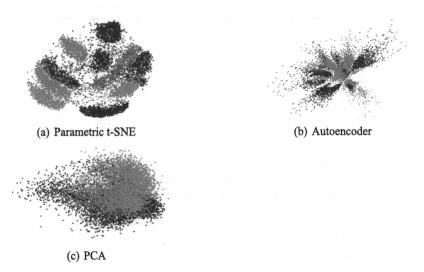

(a) Parametric t-SNE   (b) Autoencoder

(c) PCA

Fig. 3. Visualisation of normalised compressed Fisher scores derived from RBM with 5 hidden units [17].

Table 5. Accuracy of k-NN, random forest and SVM classifiers on normalised Fisher vectors derived from RBM with 5 hidden units using MNIST data set.

Classifiers	k-NN				Random Forest				SVM (Linear Kernel)			
Feature Selection Techniques	2D	10D	20D	Full D	2D	10D	20D	Full D	2D	10D	20D	Full D
MRMR	21%	59%	78%		24%	72%	85%		26%	61%	75%	
MI	16%	26%	34%		19%	36%	47%		21%	30%	40%	
Random Forest	16%	34 %	47 %	96%	19 %	48%	60%	96%	19%	39%	50%	90%
SVM-RFE	18%	30%	37%		22%	43%	53%		24%	36%	52%	
Classifiers	k-NN				Random Forest				SVM (Linear Kernel)			
Compression Techniques	2D	10D	20D	Full D	2D	10D	20D	Full D	2D	10D	20D	Full D
Parametric t-SNE	83%	93%	93.2%		88%	95%	95%		75%	94%	94.5%	
Autoencoder	75%	94%	96%		79%	94%	95%		66%	91%	89%	
PCA	26%	68%	70%	96%	30%	70%	67%	96%	36%	68 %	61%	90%
Spectral Hashing (Nbits= 16, 80, 160 for 2D, 10D and 20D respectively.)	67%	43%	38%		24%	39%	32%		14%	18%	20%	

These images are vectorised to form a 784 dimensional vector fed to the RBM's visible layer for training. A guide to initialise and optimise parameters of RBM is given by Hinton [22]. Once the model is trained generatively, it is ready for the extraction of Fisher vectors or Fisher scores for classification or retrieval applications.

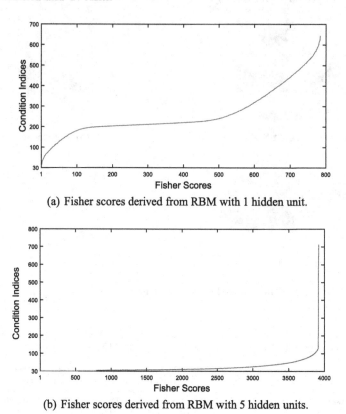

(a) Fisher scores derived from RBM with 1 hidden unit.

(b) Fisher scores derived from RBM with 5 hidden units.

Fig. 4. Estimated Belsley collinearity of Fisher vectors derived from RBM with 1 and 5 hidden units.

7.1 Experimental Setup

For experiments, we start with the extraction of Fisher vectors (FV) from compact models of RBM with 1 and 5 hidden units yielding features of size 784 and 3920 dimensions respectively. The FVs could have also been derived from a very shallow model containing thousands of hidden units as reported in [18,19], however in that case the dimensionality of the Fisher vectors scales to a magnitude of 10^6 and the model tends to over-fit resulting in no classification performance improvements [18,19]. The growth trend of deep Fisher vectors can also be observed from Table 1. We therefore constrained our compression and feature selection experiment to Fisher vectors derived from a small RBM that has shown to report the best performance on MNIST. After exploring the best condensation scheme for small architectures, we believe the same technique could be deployed for Fisher vectors derived from large models and significant improvements in storage and classification performance could be gained while avoiding overfitting. Please note that we have skipped computing Fisher scores using Eqs. 6 and 7. This is because these gradients were not found to improve the

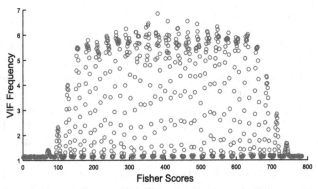

(a) Fisher scores derived from RBM with 1 hidden unit; the Fisher vector has 784 dimensions.

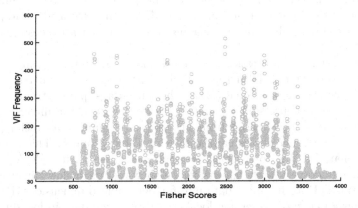

(b) Fisher scores derived from RBM with 5 hidden units; the Fisher vector has 3920 dimensions.

Fig. 5. Estimated variance inflation factor (VIF) of Fisher vectors derived from RBM with 1 and 5 hidden units.

classification accuracy of the system. In order to compute the Fisher scores, only Eq. 5 is utilised. For condensing Fisher vectors, we have applied four standard compression techniques: Principal component analysis, spectral hashing, autoencoder and parametric t-SNE, while for comparison with feature selection techniques, four popular filter and wrapper approaches are deployed, i.e. mutual information, minimum redundancy and maximum relevance, random forest and SVM-recursive feature elimination. The classification performance of both types of condensation techniques is evaluated with the help of standard classifiers: k-nearest neighbour (k-NN), random forest (RF) and support vector machines (SVM).

8   Discussion

After training RBM as a generative model, Fisher vectors are extracted using train and test examples of MNIST data set. The *train Fisher vectors* are given to the standard classifiers for training and validating model parameters, whereas the *test Fisher vectors* are used to assess their classification performance on unseen data set. In order to determine the impact of normalisation on Fisher vectors, we applied MinMax normalisation scheme to scale the data in the range [0, 1]. It is observed that compression techniques are more sensitive to normalisation as compared to the chosen feature selection methods; as a result, the compressed Fisher vectors yield better classification results after normalisation. This sensitivity is due to the nature of unnormalised data which consists of very small gradient values most of which are negative. In autoencoder and parametric t-SNE, if the data fed to the model is not positive, negative weights inhibit other neurons and the sigmoid function used in both autoencoder and parametric t-SNE saturates the gradients leading to poor classification results shown in Tables 2 and 4.

Among all the compression techniques discussed above, parametric t-SNE and autoencoder outperform the rest in terms of classification results. If one needs to store two dimensional (2D) Fisher encodings, parametric t-SNE beats the rest of the compression techniques. However, when the dimensionality of Fisher encodings is increased from 2D, the two compression techniques give comparable performances as can be seen in Tables 3 and 5. This resemblance in performance is due to the fact that parametric t-SNE first deploys autoencoder to reduce very high dimensional Fisher vectors to low dimensions and then uses t-SNE for mapping data into further reduced dimensions. Parametric t-SNE has an edge of preserving local structure of the data in low dimensional subspace using heavy tailed student t-distribution, however it is sensitive to very large data dimensions and generally uses autoencoder first for dimensionality reduction. After parametric t-SNE and autoencoder, spectral hashing and principal component analysis follow the classification league respectively. Due to the inverse relationship between Euclidian distance and Hamming affinity, spectral hashing does not guarantee to faithfully reproduce affinity between the data when the number of bits approach to infinity. It is for this reason that the classification accuracy of spectral hashing decreases as the number of code bits increases. The worst classification performance is shown by PCA, also evident from the visualisations shown in Figs. 2 and 3. PCA is not scale-invariant and mainly focuses on preserving large pairwise distances due to which it is unable to preserve the significant structure of data in low dimensional space. Also the computation of eigen vectors is infeasible for high dimensional Fisher vectors as the computation of covariance matrix becomes difficult.

On comparing the performance of feature selection schemes, the empirical results in Tables 2, 3, 4 and 5 suggest that maximum relevance and minimum redundancy (MRMR) approach outclasses the remaining feature selection methods, however it is unable to beat the discriminative performance of compression methods on normalised data. MRMR assesses individual features and selects subset of features from the top of the ranking list by assigning weights according to their degree of relevancy to the class. SVM-RFE and random forests (RF) are wrapper based feature selection schemes and therefore are trained to select subset of features that would compute best accuracy with SVM and random forest classifiers respectively. This fact can also be observed by comparing the classification accuracy yielded by all classifiers on features selected by RF and SVM-RFE. Wrappers are generally expensive to compute in comparison to filters that selects feature subsets independent of the classifier. For this reason, filters are more popular among the practitioners when the classifier to be deployed is unknown or fast and cost effective classifier is required. We have used linear SVM using stochastic gradient learning [33] as it was showing better accuracy than non-linear SVM and has the potential of scaling well with the size of the data.

In order to explore multi-collinearity in the extracted deep Fisher vectors, we utilised two diagnostic measures: (a) Variance inflation factor (VIF) and (b) Condition indices discussed in Sect. 6. Figure 4 displays the Belsley collinearity diagnostic plot for assessing the presence of near dependencies and collinearity among the dimensions of normalised Fisher vectors. The figure expresses collinearity in data by plotting condition indices of each dimension. One can observe that most of the features have extremely high condition indices that exceed the tolerance threshold, i.e. 30, exhibiting coexisting or simultaneous near dependencies in features leading to significant multi-collinearity. We also checked the variance inflation factor of our Fisher vectors for determining the presence of multi-collinearity. See Fig. 5, where the scatter plot shows that most of the features have high VIF values above the threshold 5, thus indicating the presence of multi-collinearity among dimensions.

When comparing the classification results of feature selection with compression schemes, the supremacy of compression techniques is evident on normalised data. Our results show that compressed Fisher encodings of 20 dimensions give same or better accuracy shown by full dimensional Fisher vectors, thus saving a lot more computational storage for large scale recognition problems. Moreover, the multi-collinearity diagnostics suggest that if multi-collinearity exists among dimensions then feature selection gives poor classification performance in comparison to the compression techniques. It is important to note the difference between feature selection and compression methods. Although both the techniques seek to reduce the dimensions of the data, compression methods do so by creating a new combination of attributes that project the data in a different space, whereas the feature selection methods include/exclude the data attributes without changing them and hence stay in the same data space. The presence of multi-collinearity in the data hints that there are serious problems in the attributes and small changes in the data may lead to large changes in the estimates of coefficients assisting in prediction/classification task. In order to reduce this variance in results produced due to irrelevant or redundant features, compression techniques perform better than feature selection techniques. This is because the attributes/features reduced through compression techniques are less interpretable due to space transformation and have reduced dependency on one another. This ultimately leads to better model accuracies when the data is compressed.

9   Conclusion

In this paper, we have applied different feature compression and selection techniques on Fisher vectors derived from restricted Boltzmann machine to explore their feasibility for large scale visual retrieval and classification tasks. We observed that feature compression is a better choice than feature selection if there exists multi-collinearity among features. The multi-collinearity was estimated with the help of diagnostic tests that measured the variance inflation factor and condition indices of deep Fisher vectors against the tolerance thresholds. The empirical results were shown by using condensed codes through standard classifiers such as k-NN, SVM and random forests (RF). Overall, the best classification accuracy among all feature condensation methods was yielded by parametric t-SNE and autoencoder. Parametric t-SNE and all the other compression approaches indebt their classification supremacy to MinMax normalisation method without which they are unable to compete with feature selection methods.

In future, we would like to extend this experimental framework to other large scale data sets of object recognition such as PASCAL VOC [34] and Imagenet [35]. We would also compare the algorithmic complexity of the two feature condensation approaches to explore which methods are compute and memory efficient in addition to yielding better classification performance. We believe that in applications where storage and computational resources are limited, prescribed compression techniques may prove useful for large scale object classification/retrieval tasks.

Acknowledgement. This research (SRGP: 21-402) was supported by Higher Education Commission (HEC) of Pakistan & NVIDIA (Ref.: 281400) with a valuable donation of Titan-X graphics card.

References

1. Csurka, G., Dance, C., Fan, L., Willamowski, J., Bray, C.: Visual categorization with bags of keypoints. In: Workshop on Statistical Learning in Computer Vision, ECCV, Prague (2004)
2. Everingham, M., Van Gool, L., Williams, C.K., Winn, J., Zisserman, A.: The PASCAL visual object classes (VOC) challenge. Int. J. Comput. Vis. **88**(2), 303–338 (2010)
3. Griffin, G., Holub, A., Perona, P.: Caltech-256 object category dataset. California Institute of Technology, Technical report, 7694 (2007). http://authors.library.caltech.edu/7694
4. Fei-Fei, L., Fergus, R., Perona, P.: One-shot learning of object categories. IEEE Trans. Pattern Anal. Mach. Intell. **28**(4), 594–611 (2006)

5. Farquhar, J., Szedmak, S., Meng, H., Shawe-Taylor, J.: Improving "bag-of-keypoints" image categorisation: generative models and PDF-kernels. Univ. Southampton 68 (2005)
6. Perronnin, F., Dance, C., Csurka, G., Bressan, M.: Adapted vocabularies for generic visual categorization. In: Leonardis, A., Bischof, H., Pinz, A. (eds.) ECCV 2006. LNCS, vol. 3954, pp. 464–475. Springer, Heidelberg (2006). https://doi.org/10.1007/11744085_36
7. Boureau, Y., Bach, F., LeCun, Y., Ponce, J.: Learning mid-level features for recognition. In: IEEE Computer Society Conference on Computer Vision and Pattern Recognition (2010)
8. Wang, G., Hoiem, D., Forsyth, D.: Learning image similarity from Flickr groups using stochastic intersection kernel machines. In: 2009 IEEE 12th International Conference on Computer Vision. IEEE (2009)
9. Lazebnik, S., Schmid, C., Ponce, J.: Beyond bags of features: spatial pyramid matching for recognizing natural scene categories. In: IEEE Computer Society Conference on Computer Vision and Pattern Recognition (CVPR 2006). IEEE (2006)
10. Jaakkola, T., Haussler, D.: Exploiting generative models in discriminative classifiers. In: Advances in Neural Information Processing Systems, vol. 11, pp. 487–493. MIT Press (1998)
11. Perronnin, F., Dance, C.: Fisher kernels on visual vocabularies for image categorization. In: IEEE Conference on Computer Vision and Pattern Recognition. IEEE (2007)
12. Perronnin, F., Sánchez, J., Mensink, T.: Improving the fisher kernel for large-scale image classification. In: Daniilidis, K., Maragos, P., Paragios, N. (eds.) ECCV 2010. LNCS, vol. 6314, pp. 143–156. Springer, Heidelberg (2010). https://doi.org/10.1007/978-3-642-15561-1_11
13. Chatfield, K., Lempitsky, V., Vedaldi, A., Zisserman, A.: The devil is in the details: an evaluation of recent feature encoding methods. In: Proceedings of the British Machine Vision Conference. BMVA Press (2011)
14. Perronnin, F., Larlus, D.: Fisher vectors meet neural networks: a hybrid classification architecture. In: Proceedings of the IEEE Conference on Computer Vision and Pattern Recognition, CVPR (2015)
15. Sanchez, J., Perronnin, F., Mensink, T., Verbeek, J.: Compressed fisher vectors for large-scale image classification. Rapport de recherche RR-8209, INRIA, January 2013
16. Zhang, Y., Wu, J., Cai, J.: Compact representation for image classification: to choose or to compress? In: Proceedings of the IEEE Conference on Computer Vision and Pattern Recognition. IEEE Computer Society (2014)
17. Ahmed, S., Azim, T.: Compression techniques for deep fisher vectors. In: Proceedings of the 6th International Conference on Pattern Recognition Applications and Methods (ICPRAM) (2017)
18. Azim, T., Niranjan, M.: Inducing discrimination in biologically inspired models of visual scene recognition. In: IEEE International Workshop on Machine Learning for Signal Processing (MLSP) (2013)
19. Azim, T.: Visual scene recognition with biologically relevant generative models. Ph.D. thesis, School of Electronics and Computer Science, Southampton, UK (2014)
20. Maaten, L.: Learning discriminative fisher kernels. In: Proceedings of the 28th International Conference on Machine Learning, ICML 2011, Bellevue, Washington, USA, 28 June–2 July 2011, pp. 217–224. Omnipress (2011)

21. Hinton, G.: Training products of experts by minimizing contrastive divergence. Neural Comput. **14**, 1771–1800 (2002)
22. Hinton, G.E.: A practical guide to training restricted Boltzmann machines. In: Montavon, G., Orr, G.B., Müller, K.-R. (eds.) Neural Networks: Tricks of the Trade. LNCS, vol. 7700, pp. 599–619. Springer, Heidelberg (2012). https://doi.org/10.1007/978-3-642-35289-8_32
23. Jayalakshmi, T., Santhakumaran, A.: Statistical normalization and back propagation for classification. Int. J. Comput. Theor. Eng. **3**(1), 89 (2011)
24. Weiss, Y., Torralba, A., Fergus, R.: Spectral hashing. In: Advances in Neural Information Processing Systems, NIPS (2009)
25. Hinton, G., Salakhutdinov, R.: Reducing the dimensionality of data with neural networks. Science **313**, 504–507 (2006)
26. Maaten, L.: Learning a parametric embedding by preserving local structure. In: RBM (2009)
27. Fleuret, F.: Fast binary feature selection with conditional mutual information. J. Mach. Learn. Res. **5**, 1531–1555 (2004)
28. Peng, H., Long, F., Ding, C.: Feature selection based on mutual information criteria of max-dependency, max-relevance, and min-redundancy. IEEE Trans. Pattern Anal. Mach. Intell. **27**(8), 1226–1238 (2005)
29. Guyon, I., Weston, J., Barnhill, S., Vapnik, V.: Gene selection for cancer classification using support vector machines. Mach. Learn. **46**(1), 389–422 (2002)
30. Breiman, L.: Random forests. Mach. Learn. **45**(1), 5–32 (2001)
31. Belsley, D.A.: A guide to using the collinearity diagnostics. Comput. Econ. **4**(1), 33–50 (1991)
32. LeCun, Y., Bottou, L., Bengio, Y., Haffner, P.: Gradient-based learning applied to document recognition. Proc. IEEE **86**(11), 2278–2324 (1998)
33. Bottou, L., Bousquet, O., Zurich, G.: The tradeoffs of large scale learning. In: Advances in Neural Information Processing Systems, pp. 161–168 (2008)
34. Everingham, M., Eslami, S., Van Gool, L., Williams, C., Winn, J., Zisserman, A.: The PASCAL visual object classes challenge: a retrospective. Int. J. Comput. Vis. **111**(1), 98–136 (2015)
35. Russakovsky, O., Deng, J., Su, H., Krause, J., Satheesh, S., Sean, M., Zhiheng, H., Karpathy, A., Khosla, A., Bernstein, M., Berg, A., Fei-Fei, L.: ImageNet large scale visual recognition challenge. Int. J. Comput. Vis. (IJCV) **115**(3), 211–252 (2015)

Emotion Recognition Using Neighborhood Components Analysis and ECG/HRV-Based Features

Hany Ferdinando[1,2(✉)], Tapio Seppänen[3], and Esko Alasaarela[1]

[1] Health and Wellness Measurement Research Unit,
Opto-Electronic and Measurement Technique (OPEM) Unit,
University of Oulu, Oulu, Finland
hany.ferdinando@oulu.fi
[2] Department of Electrical Engineering, Petra Christian University,
Surabaya, Indonesia
[3] Physiological Signal Team Analysis, University of Oulu, Oulu, Finland

Abstract. Previous research showed that supervised dimensionality reduction using Neighborhood Components Analysis (NCA) enhanced the performance of 3-class problem emotion recognition using ECG only where features were the statistical distribution of dominant frequencies and the first differences after applying bivariate empirical mode decomposition (BEMD). This paper explores how much NCA enhances emotion recognition using ECG-derived features, esp. standard HRV features with two difference normalization methods and statistical distribution of instantaneous frequencies and the first differences calculated using Hilbert-Huang Transform (HHT) after empirical mode decomposition (EMD) and BEMD. Results with the MAHNOB-HCI database were validated using subject-dependent and subject-independent scenarios with kNN as classifier for 3-class problem in valence and arousal. A t-test was used to assess the results with significance level 0.05. Results show that NCA enhances the performance up to 74% from the implementation without NCA with p-values close to zero in most cases. Different feature extraction methods offered different performance levels in the baseline but the NCA enhanced them such that the performances were close to each other. In most experiments use of combined standardized and normalized HRV-based features improved performance. Using NCA on this database improved the standard deviation significantly for HRV-based features under subject-independent scenario.

Keywords: NCA · Emotion recognition · ECG · HRV

1 Introduction

Previous research has reported that applying supervised dimensionality reduction (SDR) significantly enhanced the performance of emotion recognition using ECG from the MAHNOB-HCI database [1]. To be more specific, the Neighborhood Components Analysis (NCA) outperformed the Linear Discriminant Analysis (LDA) and the Maximally Collapsing Metric Learning (MCML), and the SDRs were only applied to features

© Springer International Publishing AG, part of Springer Nature 2018
M. De Marsico et al. (Eds.): ICPRAM 2017, LNCS 10857, pp. 99–113, 2018.
https://doi.org/10.1007/978-3-319-93647-5_6

resulted from one method, i.e. the statistical distribution of dominant frequencies and the first differences after applying the bivariate empirical mode decomposition (BEMD) to ECG signals, which showed its superiority in the absence of the SDR [2].

Apparently, analysis on the previous research has suffered from the number of feature extraction methods. We did not know how well the NCA can enhance the performance of the same system using features from other methods. This is the main research question addressed in this paper because conclusions based on the one feature extraction method may lead to wrong interpretation.

We applied the NCA only to features calculated using other methods, i.e. standard HRV analysis with normalization and standardization, statistical distribution of instantaneous frequency based on Hilbert-Huang Transform (HHT) after applying empirical mode decomposition (EMD) and bivariate empirical mode decomposition (BEMD), while other SDR methods were subject to future works. The results were validated under subject-dependent and subject-independent scenarios using kNN as a classifier.

The paper has been organized in the following way: the first section gives brief introduction, including a gap in the previous research. Literature studies about supervised dimensionality reduction and research in emotion recognition follows it with the main research question appears at the end of this section. The next section discusses detail methods we used in this study, including a block diagram to explain the process visually. Succeeding this section, we present experimental results along with the discussions about the findings. The last section provides conclusions and some future works.

2   Literature Studies

2.1   Supervised Dimensionality Reduction

The SDRs use classes of the samples to guide the dimensionality reduction (DR) process such that distances among points belong to the same class are decreased while increasing the distances among points belong to different class. Some proposed algorithms were, e.g. Neighborhood Components Analysis (NCA) [3], Maximally Collapsing Metric Learning [4], Large Margin Nearest Neighbor (LMNN) [5], Supervised Dimensionality Mixture Model (SDR-MM) [6], Support Vector Decomposition Machine (SVDM) [7], etc.

As was mentioned in the Introduction section that this paper focused on NCA only, the following was a brief mathematical background about the NCA, as proposed by Goldberger et al. [3]. The NCA works based on Mahalanobis distance measure

$$\|f(\mathbf{x}_1) - f(\mathbf{x}_2)\|^2 = (\mathbf{x}_1 - \mathbf{x}_2)^T \mathbf{Q}(\mathbf{x}_1 - \mathbf{x}_2) \tag{1}$$

within kNN framework, where

$$\mathbf{Q} = \mathbf{A}^T \mathbf{A} \tag{2}$$

is a positive semidefinite (PSD) learning matrix to a certain space. The algorithm aims to find the projection matrix, $\mathbf{A}$, such that the classifiers perform well in the transformed space.

By maximizing a stochastic variant of the leave-one-out (LOO) kNN score on the training data, the NCA makes no assumption about the shape of the class distribution or the boundaries between them. Since the LOO classification errors of kNN suffers from discontinuity, a differentiable cost function based on stochastic ("soft") neighbor assignments in the transformed space was introduced,

$$p_{ij} = \frac{\exp\left(-\|\mathbf{A}\mathbf{x}_i - \mathbf{A}\mathbf{x}_j\|^2\right)}{\sum_{k\neq i}\exp\left(-\|\mathbf{A}\mathbf{x}_i - \mathbf{A}\mathbf{x}_k\|^2\right)}, p_{ii} = 0 \tag{3}$$

Equation (3) assigns the probability of point i belongs to the class of selected point j, among k points as its neighbor. When point i chooses several neighbors and they might belong to different classes, total probability that point i belongs to class $C_i = \{j|c_i = c_j\}$, is defined as

$$p_i = \sum_{j\in C_i} p_{ij} \tag{4}$$

The main idea is to maximize cost function

$$f(\mathbf{A}) = \sum_i \sum_{j\in C_i} p_{ij} = \sum_i p_i \tag{5}$$

The NCA has been implemented in the *drtoolbox*, a Matlab® toolbox for dimensionality reduction [8]. Experiments in this study used this toolbox after slightly modifying the algorithm, see Sect. 3.2.

2.2 Literature Review

The MAHNOB-HCI database [9] is one of the affect recognition databases which includes ECG signals as one of the peripheral physiological signals. Other affect recognition databases which include ECG signals are RECOLA [10], Decaf [11], and Augsburg [12]. DEAP also provide signals from the heart activities but they were quantified as Heart Rate Variability (HRV) measured using Blood Volume Pulse (BVP) on finger [13]. In this paper, we use the MAHNOB-HCI database, which involved 27 subjects (11 males and 16 females) stimulated with pictures and video clips. The data includes the following synchronized signals:

- 32-channel EEG.
- Peripheral physiological signals (ECG, temperature, respiration, skin conductance).
- Face and body videos using 6 cameras.
- Eye gaze.
- Audio.

Many feature extraction methods for ECG-based emotion recognition have been proposed. HRV-based features using the standard HRV analysis were quite popular in many applications. This method requires at least 5 minutes or even hours of ECG signal to get reliable analysis [14]. There are a number of large cross-sectional studies which suggest to use other methods, some of them are non-linear point-process [15], wavelet analysis [16], Recurrent Plot [17], and empirical mode decomposition-based techniques [18].

Ferdinando et al. [19] used standard HRV analysis to get features for emotion recognition in 3-class of valence and arousal, to provide baseline for the recognition using ECG signal only from the MAHNOB-HCI database. Using SVM classifier, the achieved accuracies were 43% and 48% for valence and arousal respectively based on 10-fold cross validation. The accuracies were slightly above chance level and close enough to the ones based on all peripheral physiological signals. There was no DR applied to the acquired features even for feature selection.

Apparently, the standard HRV analysis was not suitable for ECG signals from the MAHNOB-HCI database because the signal length varies from 35–117 s. Inspired by Agrafioti et al. [18], EMD and BEMD analysis were employed [2]. Using statistical distribution of the dominant frequencies estimated from spectrogram analysis after employing BEMD analysis to ECG as features, the achieved accuracies using kNN were 56% and 60% for valence and arousal respectively based on the subject-dependent scenario. Validated under subject-independent scenario, the accuracies were 60% and 59% for valence and arousal correspondingly. Features based on statistical distribution of instantaneous frequencies estimated using Hilbert-Huang Transform (HHT) achieved less than 50% of accuracies for both valence and arousal. The only DR technique used in these experiments was feature selection.

Using SDRs implemented in *drtoolbox* [8], NCA, MCML, and LMNN, performances of the system using statistical distribution of dominant frequency after applying BEMD analysis to ECG from the same database were enhanced [1]. The NCA outperformed the other method by improving the performance significantly from 56% to 64% and from 60% to 66.1% for valence and arousal respectively in subject-dependent scenario. Under subject-independent scenario, the enhancement only worked for arousal by improving the performance from 59% to 70%.

Although the NCA showed promising results [1], the evidences reviewed in this sub-section seem to suggest evaluating how well the NCA can enhance the same system using different features, such as HRV- and HHT-based features. To our knowledge, no previous study has investigated the NCA, which was applied on the exactly same system but using features from different methods. This study can open new finding about phenomena in NCA related to different feature extraction methods as the main research question addressed in this paper.

3   Methods

Figure 1 shows the block diagram of our method. ECG signals used in these experiments were downloaded from the database server under "Selection of Emotion Elicitation" group. Sample from session #2508 was discarded as the visual inspection showed that it was corrupted, leaving 512 samples for further process. All measured signals have a synchronization pulse to separate response and baseline signals. The non-stimulations or relaxation stages are 30 s before and after stimulation part, and they must be separated to each other. We applied signal pre-processing methods suggested by Soleymani et al. [9] to remove baseline wandering and power line interference.

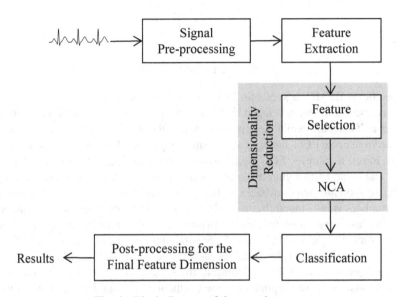

Fig. 1. Block diagram of the experiments.

3.1 Feature Extraction

We used the standard HRV analysis to extract features from both baseline and response signal as suggested by Soleymani et al. [9], i.e.

- RMS of the Successive Difference between adjacent R-R intervals (RMSSD).
- Standard Deviation of the Successive Difference between adjacent R-R intervals (SDSD).
- Standard Deviation of all NN intervals (SDNN).
- Number of pairs of adjacent NN intervals differing by more than 50 ms (NN50).
- Number of pairs of adjacent NN intervals differing by more than 20 ms (NN20).
- NN50 count divided by the total number of NN intervals (pNN50).
- NN20 count divided by the total number of NN intervals (pNN20).
- Power spectral density for very low frequency (VLF), low frequency (LF), high frequency (HF), and total power.
- Ratio of HF to LF.
- Poincaré analysis (SD1 and SD2).
- Ratio of response to baseline features.

resulted 42 features: 14 from baseline, 14 from response, and 14 from the ratios of response to baseline. We normalized them to [−1, 1] and standardized them based on mean and standard deviation to get two sets of features.

Another feature extraction method was based on the instantaneous frequency (IF) calculated using Hilbert-Huang Transform (HHT), see Eq. (6), from the intrinsic mode functions (IMFs) after either EMD or BEMD [2].

$$z(t) = d(t) + jH[d(t)]$$
$$z(t) = y(t)e^{j\theta(t)}$$
$$IF = \frac{1}{2\pi}\frac{d\theta(t)}{dt}$$

(6)

Specific to BEMD, a synthetic ECG signal, synchronized on R-wave event, was generated using a model developed by McSharry et al. [20], as the imaginary part of the complex ECG signals, while the original signal as the real part. Of note, the model only generated one cycle ECG signal as a template. By placing the template according to the R-wave event, a complex ECG signal was formed. This method was faster than generating one cycle ECG signal using the model for each detected R-wave event [2].

This method has two drawbacks, at least. First, the connection between consecutive ECG templates is not smooth but it can be minimized by adjusting the start and end of ECG template very close to zero. However, this discontinuity issue brings small problem if it is kept as small as possible. Second, the synthetic ECG may have different shape at the beginning of the signal because there is no such guarantee of getting a complete PQRST wave at the beginning of the signal, see Fig. 2. For this reason, 256 zeros were inserted at the beginning of the synthetic ECG and discarded them after the whole synthetic ECG was complete.

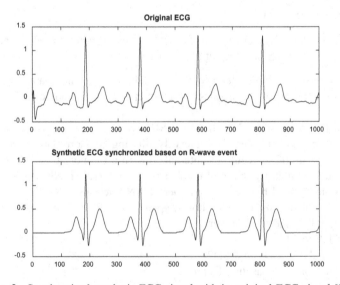

Fig. 2. Synchronized synthetic ECG signal with its original ECG signal [2].

Prior to applying EMD and BEMD, each samples was divided into 5-s segments, because both EMD and BEMD were sensitive to signal length, such they resulted 5–6 IMFs plus residue [18]. Besides, the ECGs in the MAHNOB-HCI database have different length. The IMFs from both EMD and BEMD were subject to HHT to obtain the instantaneous frequencies (IFs) from each segment. Once this process was finished, all five IFs from each segment belonging from the same ECG sample were joined to represent five IFs of that ECG signal. Following this step was to calculate 14 statistical distribution values, i.e. mean, standard deviation, median, Q1, Q3, IQR, skewness, kurtosis, percentile 2.5, percentile 10, percentile 90, percentile 97.5, maximum, and minimum, from IFs and the first differences as the features for classifier. Finally, we had another two groups of features and each group contained statistical distribution of IFs from one to five IMFs, resulting five different sets of features for each EMD and BEMD, see Table 1 for clarity. All acquired features were standardized based on mean and standard deviation (SD).

Table 1. Feature configuration prior to feature selection.

	HHT-based features after EMD	HHT-based features after BEMD
1 IMF	28 features	28 features
2 IMFs	56 features	56 features
3 IMFs	84 features	84 features
4 IMFs	112 features	112 features
5 IMFs	140 features	140 features

3.2 Dimensionality Reduction

There were two DR processes prior to classification phase, i.e. feature selection using sequential forward floating search and the NCA. Feature selection is the simplest DR technique and requires no projection matrix. It only combines the available features into a new set, with reduced dimensionality, that offers the best performance.

Using the reduced dimensionality from the previous stage, the NCA algorithm was applied to find a projection matrix able to reduce feature dimensionality while enhancing the performance. The initial projection matrix in the *drtoolbox* was set using random number such that each process produced different results and there was no guarantee that the optimum projection matrix could be acquired within single pass. For this reason, the algorithm was modified to be iterative such that it stopped when there was no improvement, validated using leave-one-out, within 200 iterations. The SDR process was applied to the selected features having dimensionality higher than the target, 2D to 9D, as in the previous study [1]. The highest possible dimensionality target, however, was 37 but it was different for each set of features. In particular, the performance analysis of the reduced dimensionality was problematic due to this limitation but we had to keep it similar to the previous one for the sake of equal methodologies [1]. The acquired projection matrices were saved for further processes.

3.3   Classifier and Validation Method

We used kNN as in the previous study [1] to make reliable comparisons before and after applying the NCA. kNN is one of the classifiers which gets benefits from DR because its reliability relies on the sample size. Using DR prior to building model with kNN saves more space for storage. Another reason related to kNN is about the computational speed. Using many samples to build a reliable model, kNN may suffer from the slow speed. This issue, however, was beyond the scope of this study because we only used 512 samples.

Results in this study were validated using subject-dependent and subject-independent scenarios. Within the subject-dependent scenario, 20% of the samples were held out for validation while the rest of them were subject to training and testing using 10-fold cross validation. The model was built based on the projection of the selected features using saved projection matrix from previous stage. The whole validation process was repeated 1000 times, with new resampling in each repetition, to accommodate the Law of Large Numbers (LLN) such that the average was close enough to the real value.

Subject-independent scenario evaluated if the features were ready for a general model where new samples were introduced to the classifier for recognition. Samples from one subject were excluded from building the model and used them to test the model. This process continued for all subjects and the reported performance was the average over all exclusion processes. We called this validation as Leave-One-Subject-Out (LOSO) validation.

3.4   Post-processing for the Final Feature Dimension

Validations tests were designed so that they produced classification accuracy with several dimensionalities to select the best one. However, small differences between the accuracies may not be statistically significant. Specifically, there can be two accuracies close to each other while the feature vector dimensions are different. It would make sense to choose the one that has a lower dimension. The following procedure was therefore used to choose the final feature vector dimension:

1. Find the best accuracy (namely, A1).
2. If the best accuracy is occurred at the lowest dimensionality, then the best result is found (best result = A1).
3. Otherwise, find the second-best accuracy (namely, A2) from the lower dimensionality and compare A1 to A2 using t-test with significance level 0.05.
4. If the difference is statistically significant, then the best results is found (best result = A1).
5. If the difference is not statistically significant, then the second-best turns to the best accuracy. Repeat process from step 2 until it reaches the lowest dimensionality.

4   Results and Discussions

We experimented with four sets of features extracted using different methods and then compare the results side-by-side. We provided the baseline performances for each set of features, evaluated the difference using t-test with significance level 0.05, and calculated the improvement in percent to answer the main research question in this paper. Results from the previous study [1] were also presented.

4.1   HRV-Based Features

Table 2 shows the experiment results using standardized HRV-based features under subject-dependent scenario. The highest performances were 60% and 46% for valence 4D and arousal 3D respectively. Using the post-processing procedure for the final feature dimension in Sect. 3, the second highest performance in valence with lower dimensionality was at 2D and p-value as the result of significance test was close to zero, indicating that the difference was significant such that 60.0 ± 4.4 was the best result occurred at the lowest dimensionality. Applying the same rules for arousal, we compared the one at 3D to 2D using t-test and resulted a very small p-value specifying that 46.0 ± 4.1 in 3D was better than the other. Performances improved about 17% and 6% for valence and arousal respectively.

Table 2. Results for standardized HRV-based features for subject-dependent scenario from each dimensionality.

	Baseline	2D	3D	4D	5D	6D
Valence	51.2 ± 4.2	57.8 ± 4.3	57.2 ± 4.2	**60.0 ± 4.4**	58.8 ± 4.4	59.4 ± 4.3
Arousal	43.3 ± 4.2	45.2 ± 4.3	**46.0 ± 4.1**	44.2 ± 4.3	43.8 ± 4.1	44.2 ± 4.4

Results from standardized HRV-based features within subject-independent scenario are presented in Table 3. For valence, the highest accuracy was at 4D but significance test against the one at 3D gave p-value 0.079 indicating that the difference was not significant such that result at 3D became the best one. Next, we compared result at 3D to 2D, the second highest result, and found that the difference was significant, brought 61.4 ± 4.0 at 3D as the best result. For arousal, comparing the highest performance at 3D, 42.8 ± 4.0, to the second highest one at 2D, 41.7 ± 4.2, emerged p-value 0.0866, such that result from 2D was chosen as the best result. Although the NCA worked well in valence, no evidence was found for improvement in arousal.

Table 3. Results for standardized HRV-based features for subject-independent scenario from each dimensionality.

	Baseline	2D	3D	4D	5D	6D
Valence	54.1 ± 11.3	55.2 ± 4.8	**61.4 ± 4.0**	62.8 ± 4.9	60.3 ± 4.6	61.5 ± 4.3
Arousal	44.5 ± 8.0	**41.7 ± 4.2**	42.8 ± 4.0	36.9 ± 4.2	36.9 ± 4.1	42.0 ± 4.7

The same procedures were applied to all sets of features and the results were summarized in Tables 4, 5, 6 and 7 for valence and arousal within both scenarios. From the third row of those tables, it was evident that NCA improved all performances significantly except for arousal in subject-independent scenario and it was even lower than its baseline. If we compared results from arousal for both scenarios, the second column of Tables 5 and 7, they were either similar or even worse than the ones in [19] although the later used neither feature selection nor NCA. It was also shown here that the improvements from the baseline was somehow small.

If we now turn to experiment with normalized HRV-based feature, the third column of Tables 4, 5, 6 and 7, it was apparent that NCA improved all performances significantly as shown by p-values at the last row. Compare to standardized HRV-based features, improvements for normalized HRV-based features were considerably better than the other.

Surprisingly, experiments within subject-independent scenario showed that the SD reduced around 50% after applying NCA, indicating higher consistency among the repetitions than the ones in baselines, but not in subject-dependent scenario. These were unexpected as validation using this scenario usually resulted high variation.

These experiments also presented the fact that whether valence had higher accuracy than arousal and the other way around depended on the normalization method. This finding brought an idea to combine standardized and normalized HRV-based feature, select the most discriminant features, and then apply NCA to evaluate if this combination offers more powerful features than working individually. By combining these two sets of features, result for arousal under subject-independent scenario looked promising, see the fourth column of Tables 4, 5, 6 and 7. Besides, the accuracies were even better than the ones when both sets of features worked individually. The selected features in this scheme were from both parties showing that combining these two set of features was a choice.

4.2 HHT-Based Features

Performances for recognition in valence and arousal using HHT-based feature after EMD analysis under both scenarios are presented in the fifth column of Tables 4, 5, 6 and 7. Applying NCA to HHT-based feature after EMD analysis enhanced the performance for both valence and arousal in both scenarios significantly, indicated by p-values, with the largest enhancement occurred in arousal under subject-independent scenario, see Table 7.

The sixth column of Tables 4, 5, 6 and 7 displays the summary of experiments using HHT-based feature after BEMD analysis. The improvements were large and enhanced the performance significantly as well. Of note, the baselines of HHT-based feature after BEMD were mostly the smallest among all experiments such that it offered the largest improvement.

4.3 Summary of the Experiments

Tables 4, 5, 6 and 7 present comparisons side-by-side for each emotional label under subject-dependent and subject-independent scenarios. Generally, the NCA could

Table 4. Compare side-by-side for valence in subject-dependent scenario.

	Standardized HRV-based features	Normalized HRV-based features	Combined standardized and normalized HRV-based feature	HHT-based features after EMD analysis	HHT-based features after BEMD analysis	Spectrogram-based features with BEMD analysis
Baseline	51.2 ± 4.2	48.6 ± 4.5	45.9 ± 4.4	50.5 ± 4.5	45.8 ± 4.3	55.8 ± 7.3
After NCA	60.0 ± 4.4	57.6 ± 4.4	67.1 ± 4.4	50.5 ± 4.5	**68.6 ± 4.4**	64.1 ± 7.4
Improvement	17%	19%	46%	19%	50%	15%
p-value	~0	~0	~0	~0	~0	~0

Table 5. Compare side-by-side for arousal in subject-dependent scenario.

	Standardized HRV-based features	Normalized HRV-based features	Combined standardized and normalized HRV-based feature	HHT-based features after EMD analysis	HHT-based features after BEMD analysis	Spectrogram-based features with BEMD analysis
Baseline	43.3 ± 4.2	53.3 ± 4.6	55.8 ± 4.6	50.2 ± 4.5	47.3 ± 4.1	59.7 ± 7.0
After NCA	46.0 ± 4.1	61.8 ± 4.5	**70.7 ± 4.3**	66.6 ± 4.5	64.1 ± 4.3	66.1 ± 7.4
Improvement	6%	16%	27%	21%	36%	11%
p-value	~0	~0	~0	~0	~0	~0

Table 6. Compare side-by-side for valence in subject-independent scenario.

	Standardized HRV-based features	Normalized HRV-based features	Combined standardized and normalized HRV-based feature	HHT-based features after EMD analysis	HHT-based features after BEMD analysis	Spectrogram-based features with BEMD analysis
Baseline	54.1 ± 11.3	50.0 ± 11.4	44.7 ± 10.4	48.3 ± 10.0	43.0 ± 9.5	59.2 ± 11.4
After NCA	61.4 ± 4.0	59.7 ± 4.5	**70.7 ± 4.9**	61.2 ± 11.9	62.8 ± 12.1	61.7 ± 14.1
Improvement	13%	19%	58%	27%	46%	4%
p-value	~0	~0	~0	~0	~0	0.1873

Table 7. Compare side-by-side for arousal in subject-independent scenario.

	Standardized HRV-based features	Normalized HRV-based features	Combined standardized and normalized HRV-based feature	HHT-based features after EMD analysis	HHT-based features after BEMD analysis	Spectrogram-based features with BEMD analysis
Baseline	44.5 ± 8.0	53.6 ± 9.7	55.7 ± 11.0	47.0 ± 13.5	44.1 ± 10.9	58.7 ± 9.1
After NCA	41.7 ± 4.2	62.7 ± 4.0	**73.5 ± 4.4**	65.7 ± 12.6	63.8 ± 11.9	69.6 ± 12.4
Improvement	-6%	17%	32%	40%	45%	18%
p-value	0.0432	~0	~0	~0	~0	~0

improve the performances in both valence and arousal within both scenarios. An exception occurred for valence in subject-independent scenario using spectrogram-based features after BEMD [1], see the most right column of Table 6, and arousal in the same scenario using standardized feature, see the second column of Table 7, as the NCA failed to make it.

Although the baselines for different set of features had large differences, the results after applying NCA were quite good by neglecting the exceptional cases above. The lowest accuracies before NCA corresponded to the highest improvements and vice versa. These facts were interesting as the NCA could make the end results almost close to each other no matter the feature extraction method applied to ECG signals. To our knowledge, this interesting phenomenon has not been exposed before.

Related to computational cost, HRV-based feature extraction offered the lightest one with moderate performances after feature selection process and good results after applying NCA. On the other hand, a method utilizing spectrogram after BEMD had high cost because of the spectrogram analysis parameters, i.e. seven values of window size and nine values of overlap parameters, and process related to BEMD [2]. Moreover, the feature selection process and NCA must search for all parameter combinations. Furthermore, the end results from spectrogram-based feature after BEMD analysis could not beat the ones from combined standardized and normalized HRV-based features although the former won the competition on the baseline.

Experiments under subject-independent scenario expect higher variances as the classifiers never learn the structure from training data [1, 2]. However, NCA was able to successfully reduce the SDs significantly for HRV-based features. Even more interesting, the NCA lowered them such that the values were close to the other scenario. There was no such improvement from the other feature extraction methods. This finding was also interesting but it needs more studies with other databases and feature extraction methods, and is left for future work.

5 Conclusions

Enhancements of ECG-based emotion recognition on the MAHNOB-HCI database, processed by several feature extraction methods, using NCA were presented. Generally, NCA could successfully enhance the performance on this database significantly and provided new baselines. Results using combined standardized and normalized HRV-based features were superior, except for valence in subject-dependent scenario. Although spectrogram-based features after BEMD analysis outperformed the other feature extraction methods when NCA was not applied [2], the results were completely different after applying NCA as shown in Tables 4, 5, 6 and 7.

Different feature extraction methods had different classifier performances but the NCA could make the results from different methods closer to each other. This fact was interesting to note because so far feature extraction methods were very critical. However, this observation needs more elaboration with other databases and feature extraction methods.

Spectrogram-based features after BEMD analysis had a heavy computational cost and the performances after NCA were not as good as in the baseline. On the other hand,

HRV-based features had a light computational cost but offered better results after NCA. Having higher baseline accuracy brought no guarantee that applying NCA would result in as good improvement as with lower baseline levels.

The NCA reduced the SDs around 50% from the baseline on experiments using HRV-based features under subject-independent scenario. To our knowledge, such results have not been explored before in many experiments using NCA. However, confirmation using other databases but MAHNOB-HCI database remains as future work.

References

1. Ferdinando, H., Seppänen, T., Alasaarela, E.: Enhancing emotion recognition from ECG signals using supervised dimensionality reduction. In: Proceedings of the 6th International Conference on Pattern Recognition Applications and Methods - Volume 1: ICPRAM, pp. 112–118. Scitepress, Porto (2017)
2. Ferdinando, H., Seppänen, T., Alasaarela, E.: Comparing features from ECG pattern and HRV analysis for emotion recognition system. In: 2016 IEEE Conference on Computational Intelligence in Bioinformatics and Computational Biology (CIBCB), pp. 1–6. IEEE. Chiang Mai (2016)
3. Goldberger, J., Roweis, S., Hinton, G., Salakhutdinov, R.: Neighbourhood components analysis. Adv. Neural. Inf. Process. Syst. **17**, 1–8 (2004)
4. Globerson, A., Roweis, S.: Metric learning by collapsing classes. Neural Inf. Process. Syst. **18**, 451–458 (2005)
5. Weinberger, K.Q., Saul, L.K.: Distance metric learning for large margin nearest neighbor classification. J. Mach. Learn. Res. **10**, 207–244 (2009)
6. Orlitsky, A.: Supervised dimensionality reduction using mixture models. In: Proceedings of the 22nd International Conference on Machine Learning (ICML), pp. 768–775. ACM, Bonn (2005)
7. Pereira, F., Gordon, G.: The support vector decomposition machine. In: Proceedings of the 23rd International Conference on Machine Learning (ICML), pp. 689–696. ACM, Pittsburgh (2006)
8. van der Maaten, L.: Matlab Toolbox for Dimensionality Reduction. https://lvdmaaten.github.io/drtoolbox/
9. Soleymani, M., Lichtenauer, J., Pun, T., Pantic, M.: A multimodal database for affect recognition and implicit tagging. IEEE Trans. Affect. Comput. **3**, 42–55 (2012)
10. Ringeval, F., Sonderegger, A., Sauer, J., Lalanne, D.: Introducing the RECOLA multimodal corpus of remote collaborative and affective interactions. In: 2013 10th IEEE International Conference and Workshops on Automatic Face and Gesture Recognition FG 2013 (2013)
11. Abadi, M.K., Subramanian, R., Kia, S.M., Avesani, P., Patras, I., Sebe, N.: DECAF: MEG-based multimodal database for decoding affective physiological responses. IEEE Trans. Affect. Comput. **6**, 209–222 (2015)
12. Wagner, J., Kim, J., Andre, E.: From physiological signals to emotions: implementing and comparing selected methods for feature extraction and classification. In: 2005 IEEE International Conference on Multimedia and Expo, pp. 940–943. IEEE (2005)
13. Koelstra, S., Muhl, C., Soleymani, M., Lee, J.-S., Yazdani, A., Ebrahimi, T., Pun, T., Nijholt, A., Patras, I.: DEAP: a database for emotion analysis using physiological signals. IEEE Trans. Affect. Comput. **3**, 18–31 (2012)

14. Task-force: Heart rate variability. Standards of measurement, physiological interpretation, and clinical use. Task Force of the European Society of Cardiology and the North American Society of Pacing and Electrophysiology. Eur. Heart J. **17**, 354–381 (1996)

15. Valenza, G., Citi, L., Lanatá, A., Scilingo, E.P., Barbieri, R.: Revealing real-time emotional responses: a personalized assessment based on heartbeat dynamics. Sci. Rep. **4**, 4998 (2014)

16. Konar, A., Chakraborty, A.: Emotion Recognition: A Pattern Analysis Approach. Wiley, Hoboken (2015)

17. Valenza, G., Lanata, A., Scilingo, E.P.: The role of nonlinear dynamics in affective valence and arousal recognition. IEEE Trans. Affect. Comput. **3**, 237–249 (2012)

18. Agrafioti, F., Hatzinakos, D., Anderson, A.K.: ECG pattern analysis for emotion detection. IEEE Trans. Affect. Comput. **3**, 102–115 (2012)

19. Ferdinando, H., Ye, L., Seppänen, T., Alasaarela, E.: Emotion recognition by heart rate variability. Aust. J. Basic Appl. Sci. **8**, 50–55 (2014)

20. McSharry, P.E., Clifford, G.D., Tarassenko, L., Smith, L.A.: A dynamical model for generating synthetic electrocardiogram signals. IEEE Trans. Biomed. Eng. **50**, 289–294 (2003)

A Conversive Hidden Non-Markovian Model Approach for 2D and 3D Online Movement Trajectory Verification

Tim Dittmar[(✉)], Claudia Krull, and Graham Horton

Faculty of Computer Science, Otto-von-Guericke-University Magdeburg,
39106 Magdeburg, Germany
tim.dittmar@ovgu.de

Abstract. A novel approach for stochastically modelling movement trajectories is presented that has already been implemented and evaluated for classification scenarios in previous research and in this article its applicability to verification scenarios is analysed. The models are based on Conversive Hidden non-Markovian Models that are especially suited to mimic temporal dynamics of time series. In contrast to the popular Hidden Markov Models (HMM) and the dynamic time warping (DTW) method, timestamp information of the data is an integral part. A verification system is presented that create trajectory models from several examples and its verification performance is deduced from experiments on different data sets including signatures, doodles, pseudo-signatures and hand gestures recorded with a Kinect. The results are compared to other publications and they reveal that the developed system already performs similar to a general DTW approach, but expectedly does not yet reach the quality of specialized HMM systems. It is also shown that the system can be applied to three dimensional data and further possibilities to improve the results are discussed.

Keywords: Online signature verification
Conversive Hidden non-Markovian Model · DTW · HMM
Movement trajectories · Kinect

1 Introduction

Human movements are a natural way of interacting with our environment, such as other beings or objects. Human computer interaction (HCI) is concerned with how humans can efficiently and effectively interact with computers. Apart from that, computer based analysis of movements can also be relevant for security, forensic analysis, or sport science. When analysing designated movements, usually the shape of the path of a movement (trajectory) and its temporal dynamic are of interest. However, due to variations between repeated execution of the same movement or between executions by different people, verification and classification can prove difficult.

© Springer International Publishing AG, part of Springer Nature 2018
M. De Marsico et al. (Eds.): ICPRAM 2017, LNCS 10857, pp. 114–131, 2018.
https://doi.org/10.1007/978-3-319-93647-5_7

The modelling paradigm of Conversive Hidden non-Markovian Models (CHnMM) can capture the temporal and spatial properties of a movement trajectory, as well as the deviations that can occur between different executions. In previous research (see Sect. 2.1), we have shown that CHnMM can efficiently model and classify movement trajectories of touch gestures. CHnMM can be automatically created from training examples and show a promising performance in this area. In the current article, we present further research on the applicability of CHnMM in verification tasks.

One common application involving movement trajectories is authentication via signature verification. To be able to compare CHnMM with standard signature verification methods, we used publicly available databases containing sufficient data suitable for the task. Besides actual online signature data, we also used finger drawn doodles and pseudo signatures to assess our methods performance. All of these contain spatial as well as temporal information on the movement trajectories, which is sufficient for our method. Since CHnMM are not specifically tailored for signature verification, but generally applicably to spatio-temporal trajectories, we do not expect to outperform specialized systems. Nevertheless, the goal of the current research is to show that CHnMM are applicable for movement trajectory verification tasks using real world application data.

We have designed CHnMM to capture not just the shape but also the temporal dynamics of a movement. Therefore we expect them to be able to distinguish trajectories that are similar in shape but differ in temporal execution. This trait could be beneficial when deciding whether a signature is valid. An actual forgery attempt may be able to mimic a trajectories shape, but will probably exhibit different temporal dynamics.

2 Related Work

2.1 Previous Work

Hidden non-Markovian Models (HnMM) as an extension of Hidden Markov Models (HMM) [16] have been developed by [11] and allow more realistic modelling of systems, allowing for multiple concurrent non-Markovian processes. A state space-based solution method for behaviour reconstruction solving the Evaluation and Decoding problem has been developed, which is computationally very demanding. The subclass of Conversive HnMM (CHnMM) [2] slightly reduced the modelling power but significantly improved the efficiency of the behaviour reconstruction algorithms.

In two studies we evaluated the applicability of CHnMM to gesture and pattern recognition, as one major application are of HMM. [1] evaluated WiiMote movement classification and [3] tested touch gesture recognition. Both studies revealed that CHnMM can outperform HMM in terms of recognition accuracy in cases when the shape of the gestures is not the discriminating factor but its temporal dynamics.

However, in both studies, the gesture models were created manually from captured movement executions, which basically prevents the practical applicability of the approach. In [4] an automatic model creation approach has been developed that covers general movement trajectories that spatially and temporally behave similar on each repetition. This has been implemented and tested on touch gesture recognition tasks with promising results. In this work the developed concept is utilized, adapted and applied to verification problems to evaluate its potential in this field of application.

2.2   Related Work

Online signature verification is already a well known research area with numerous methods and techniques being utilized which can be separated into two main categories of systems: 'feature-based' and 'function-based' [13]. 'Feature-based' systems calculate global features from the input data to do the verification while 'function-based' systems operate on the time-discrete function describing the pen or finger movement trajectory. The CHnMM system is a 'function-based' one and two main representatives in this category are HMM and DTW based systems of which plenty exist.

One instance is the work by Fierrez et al. [7] where a HMM based system is employed, extracting several different features from online signatures (from MCYT database) to train continuous HMM from examples where each trained HMM is a representation of a certain signature. Similarly, Muramatsu and Matsumoto [14] learned discrete HMM only using the quantized direction angle as a feature to model Chinese online signatures. The training process for HMM requires a significant amount of time to generate the model and compared to DTW it also tends to require more training examples to produce models of good quality [7]. The computation of the verification score however, is comparably fast.

A very common technique for online verification is DTW that calculates the distance between two time series with different length. Hence, it can be employed for a template matching approach. Kholmatov and Yanikoglu [10] for example developed a DTW based online verification system that won the First International Signature Verification Competition even without using further information like pressure, azimuth or elevation of the stylus. Other examples of DTW based verification systems are described by Faundez-Zanuy [6] and Martinez-Diaz et al. [12]. In [12] the DTW method was applied on finger-drawn doodles and pseudo-signatures recorded on a mobile touch device. In contrast to the HMM method, DTW requires all the training examples to be stored as templates. To verify an input a DTW distance score is determined for each available template and if a certain threshold is exceeded the input is considered to be valid. This method is also applied on three dimensional data: Tian et al. [17] utilized the DTW approach to verify signatures that were written in the air and recorded via a Kinect device.

Although the temporal dynamics are essential to verify a signature, neither HMM nor DTW utilize any time information in the calculations. They assume a regular time series like a fix frequency from a recording device. Both methods could unveil problems in cases where this frequency changes for example because of different recording devices. CHnMM explicitly require the timestamp of each observation but are not bound to regular or fix-frequency signals.

3   The CHnMM Verification System

In this section the extended CHnMM-based verification system for spatio-temporal movement trajectories and its most important elements are explained. The idea, methods and algorithms of the system are still similar to the two dimensional system explained in [5] but remarks about the adaptations to enable the processing of three dimensional trajectories are included.

3.1   CHnMM - Formal Definition

For a better understanding of the following explanations the formal definition of a CHnMM is given, introducing terms and nomenclature.

Certain elements of CHnMM are similar to the elements of HMM, namely:

- $S = (s_1, \ldots, s_N)$: set of N discrete states
- $V = (v_1, \ldots, v_M)$: set of M output symbols
- $\Pi = (\pi_1, \ldots, \pi_N)$: initial probability vector
- A: $N x N$ transition matrix with elements a_{ij} being more complex.

In addition to these similar elements, CHnMM also contain of a set of K transitions $TR = \{tr_1, tr_2, \ldots, tr_K\}$ that describe state change behaviour and thereby model behaviour. Each a_{ij} in A is either element of TR or $\emptyset$ if there is no transition between state s_i and s_j. Properties of the state change are described by the transition tr_i, which is a tuple of three elements $(dist, b(v), aging)$.

As a part of the element a_{ij} the continuous probability distribution $dist$ describes the duration from state s_i becoming active until the discrete state change to state s_j occurs. The function $b(v)$ determines the probability that the symbol v is emitted when the state change is happening and therefore $b(v)$ is semantically equivalent to the output probabilities described in matrix B of HMM, except that symbols are associated to transitions instead of states. In case of multiple outgoing transitions from state s_i the boolean value $aging$ determines if the elapsed time since the transition became active (i.e. the state s_i became active) is saved ($aging = true$) or not ($aging = false$) when the transition becomes disabled before firing. This value is not of further interest in this paper and is always considered to be set to false.

In conclusion, a complete CHnMM λ is formally described with the elements as a tuple, i.e. $\lambda = (S, V, A, TR, \Pi)$.

3.2   Trajectory Model Structure

The chosen model structure for the verification system is driven by the idea to split the stochastic process into its spatial and temporal stochastics, where the stochastic process is considered to be the body movement that is always generating rather similar trajectories. This separation facilitates the automatic CHnMM creation by utilizing the spatial information of example trajectories to define the CHnMM states S, the output symbols V and their output probabilities $tr.b(v)$. With that given as a base, the temporal stochastics of the body movement to be modelled are extracted from examples and encoded in the transitions of the CHnMM via the temporal probability distributions $tr.dist$.

To represent the spatial stochastics of the process, the so called *StrokeMap* was introduced, which consists of regions each trajectory path will reach successively. For two dimensional trajectories, these regions are circular, while for three dimensional trajectories they are extended to spheres. A visualization of the general modelling idea and approach is given in Fig. 1. It shows two example trajectories (as points represented by blue circles), created by the stochastic process/body movement, which are used to automatically create the *StrokeMap*. Thereupon, the *StrokeMap* serves as the base for the structure and layout of the CHnMM, whose transition time distributions are estimated from the trajectory examples. The following two sections describe and explain the details of the automatic generation of the CHnMM trajectory models.

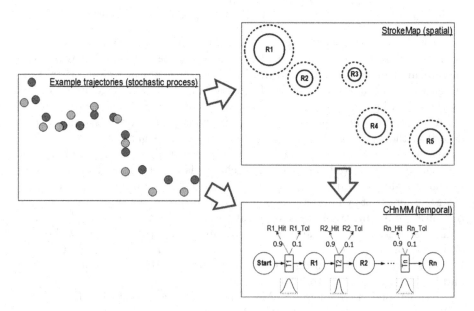

Fig. 1. The approach: split the stochastic process given by example trajectories into its spatial and temporal stochastics [5]. (Color figure online)

3.3   Creating the StrokeMap

The *StrokeMap* is represented as an ordered set of regions ($SM = \{R_1, \ldots, R_n\}$) resembling the spatial locations each trajectory has to pass through successively, if it resembles the body movement to be modelled. By defining probable locations where the trajectory points will occur, the spatial stochastics are integrated into the model. Each region consists of its position (2D or 3D), its radius and its tolerance radius ($R = (x, y, z, r, r_{tol})$) and either forms a circle or a sphere depending on the dimensionality of the trajectories. It would also be possible to employ other shapes like ellipses and ellipsoids as region shapes, which will be tested in future research. The automatic generation of regions is based on the set of example trajectories $I = \{trj_1, \ldots, trj_n\}$ that is provided as an input, with each trajectory being a chronologically ordered sequence of recorded trajectory points, each holding information about the position at a certain point in time ($trj = ((x_1, y_1, z_1, t_1), \ldots, (x_n, y_n, z_n, t_n))$).

A more precise and formal description of the *StrokeMap* generation process is given by Algorithm 1, that explains how regions R_1 to R_n are determined. The first step is the interpolation of each trajectory in I to approximate a continuous movement path that has a position at every point in time of the trajectory. Subsequently, for each trajectory, a fixed number of spatially equidistant points is sampled from its interpolated path determined by the parameter *nRegions*. Hence, the arc distance between the points Δs_{trj} depends on this parameter and the arc length of the complete interpolated path.

$$\forall trj \in I :$$
$$Int_{trj}(s) = Interpolation(trj)$$
$$\Delta s_{trj} = \frac{Length(Int_{trj})}{nRegions}$$
$$\forall i \in \mathbb{N}, 1 \leq i \leq nRegions :$$
$$RP_i = \{rp_{i,trj} \mid Int_{trj}(\Delta s_{trj} * i)\}$$
$$D_i = \{\Delta t \mid rp_{i,trj}.t - rp_{i-1,trj}.t\}$$
$$R_i = CreateRegion(RP_i, minRadius)$$
$$R_i.r_{tol} = R_i.r * toleranceFactor$$

Algorithm 1. StrokeMap generation [5].

Determined by their region index, the sampled points are grouped together, forming the set of region points RP_i. Each group is the base for the creation of a specific region R_i of the *StrokeMap*. This process is represented by the *CreateRegion* function that determines the radius and the position of a minimal circular or spherical region containing every point in the given set RP_i. To avoid small regions, e.g. due to a limited number of examples, the parameter *minRadius* is utilized, defining the minimal radius of regions returned by *CreateRegion*.

It can be expected that the path of unknown or new executions of the trajectory may not lie within the calculated regions but close, the parameter *toleranceFactor* is implemented to define a bigger tolerance region by multiplying the factor with the original region radius, thus creating a bigger circle or sphere.

Although not necessary for the *StrokeMap* itself, the set D_i containing the times needed to travel the Δs_{trj} distance from region R_{i-1} to R_i is already calculated, as it will be needed in the CHnMM creation process explained in the next section.

3.4   Creating the CHnMM

The automatic construction of the final CHnMM is formally described in Algorithm 2. As mentioned before, the layout and structure of the CHnMM is mainly defined by the *StrokeMap*. This becomes obvious considering the fact that the elements that define the structure S, V, A are already determined by knowing the parameter *nRegions*. Each state corresponds to an region of the *StrokeMap* and represents a certain phase of the movement. Due to their chronological order, a linear topology is employed to connect the states with transitions as it is similarly known from HMM [8]. A graphical representation of the CHnMM structure is given in Fig. 1.

With the layout being defined, only the transitions tr_i of the CHnMM are left. For the output probabilities the parameter *hitProbability* is utilized, specifying the probability that the Ri_Hit symbol is generated by a trajectory. Semantically, this indicates that the according sampling point rp_i of a given trajectory lies within the circular or spherical core region. If only the tolerance region is reached, the symbol Ri_Tol is emitted, which is penalized by using a smaller probability. As a consequence, *hitProbability* must be greater than 0.5.

$$S = \{Start, R_1, \ldots, R_n\}$$
$$V = \{\text{R1\_Hit}, \text{R1\_Tol}, \ldots, \text{Rn\_Hit}, \text{Rn\_Tol}\}, \ n = nRegions$$
$$A = TR^{nRegions \times nRegions}, a_{ij} = \begin{cases} tr_j & \text{if } j = i+1 \\ \emptyset & \text{otherwise} \end{cases}$$
$$\forall i \in \mathbb{N}, 1 \leq i \leq nRegions:$$
$$tr_i.b(\text{Ri\_Hit}) = hitProbability$$
$$tr_i.b(\text{Ri\_Tol}) = 1 - hitProbability$$
$$tr_i.aging = false$$
$$tr_i.dist = CreateDistribution(D_i, distType)$$

Algorithm 2. CHnMM generation [5].

For the probability distribution of a transition $tr_i.dist$, that defines the temporal behaviour, the set D_i from the *StrokeMap* creation is passed to the *CreateDistribution* function that estimates a fitting distribution according to the given *distType*.

3.5   Trajectory Verification

After a trajectory model consisting of the *StrokeMap* and the CHnMM has been created, it can be used to verify unknown trajectory examples. Therefore, the evaluation task, which is known from HMM systems, needs to be solved. Formally this means to calculate $P(O|\lambda)$ given a symbol trace $O = (o_1, \ldots, o_T)$ and a CHnMM λ. The symbol trace O is generated from the unknown input trajectory by using the point sampling method from Sect. 3.3. If a point lies within its corresponding *StrokeMap* region either Ri_Hit or Ri_Tol is emitted as an observation o_i at the interpolated time of the sample point. If there is a single sample point that does not lie within its region, the result for $P(O|\lambda)$ is 0, otherwise the probability that the model λ created the trace O is calculated according to the evaluation algorithm presented in [2].

If the result is 0, the input is assumed to be invalid, which for example happens if the trajectory does not pass a tolerance region or if the time needed from region to region does not fit with the probability distribution. Instead of 0 a threshold value could be introduced, which is discussed in Sect. 4.3 and which is common for most authentication systems.

4   Experiments

4.1   Databases

For the experiments different real world datasets containing trajectory data were utilized. The majority of the data was collected and intended for biometric authentication purposes. It is interesting to evaluate these with the developed CHnMM authentication system, since the data was recorded with different devices and by a sufficient number of persons.

MCYT. A typical real world biometric authentication application is the usage of signatures to verify that a person is who he/she claims to be. A database containing many real world online signatures is the so called MCYT (Ministerio de Ciencia y Tecnología) bimodal biometric database [15] whose intention is to represent a statistical significant part of a large scale population. It ideally suits the purpose of evaluating the CHnMM authentication system, because its performance can be measured and compared to systems that also utilized this dataset. It is kindly provided by Biometric Recognition Group - ATVS of the Universidad Autonoma de Madrid.

The dataset contains signatures of 100 participants of which each created 25 genuine examples of his/her signature using a WACOM INTUOS A6 USB pen tablet. With a 100 Hz frequency it records: x-y coordinates, the applied pen pressure and the azimuth and altitude angle of the pen relative to the tablet. Additionally, the dataset also contains 25 forgeries per participant. These were created by showing a static image of the genuine signature to five other users who tried to replicate it five times. For three different participants a few signature recordings are presented in Fig. 2.

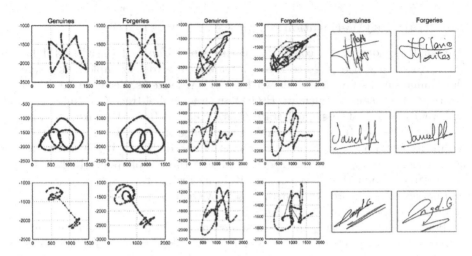

Fig. 2. Genuine and forgery examples from the Doodle (left), PseudoSignatures (middle) and MCYT (right) database corpora [5].

To facilitate the processing of the data by the CHnMM authentication system, a synthetic timestamp is added to the recorded data by increasing the timestamp by 10ms for each successive feature vector. The CHnMM system only utilizes the x-y coordinates and the timestamp for the verification, because it was designed for general movement trajectories and not device or application specific data.

DooDB. While the previous trajectories were created with a pen tablet, the DooDB contains trajectory data of 2-dimensional finger movements. The database was generated by Martinez-Diaz et al. [12] and is made publicly available by the ATVS group. There are two different corpora, namely Doodles and Pseudo-signatures and for both an HTC Touch HD mobile phone (5 × 8.5 cm screen) was employed to record single finger movements on the touch surface. For their creation 100 participants performed 30 genuine examples and 20 forgeries in both corpora, which vary in what the participants were asked to draw: In Doodles participants draw a doodle that they would use regularly as a graphical password for authentication scenarios and in Pseudo-signatures a simplified version of their signature (shorter or only initials to fit the screen) was requested.

Similarly in both corpora the x-y coordinates and a time interval for the elapsed time since the previous touchpoint (most commonly 10 ms due to 100 Hz device frequency) are recorded. In case of the finger losing contact with the touch surface no data is recorded as opposed to the data in MCYT in case the pen is just hovering over the tablet. If erroneous recordings, i.e. 0, 0 coordinates, occur they are ignored for the trajectory. However, their elapsed time information is still considered to determine the correct timestamps for successive touchpoints.

KinectDB. This database was specifically created by students of our department (Lehrstuhl für Simulation) to evaluate recognition and verification methods on 3D trajectories with a focus on temporal dynamics. The database consists of 3D hand trajectories of people performing certain gestures with their right hand that were recorded with a Microsoft Kinect (v1) device. A focus was set on a gesture set that also includes gestures that exhibit the same shape but different temporal dynamics. This gesture set consisting of nine different gestures is visualized in Fig. 3, where colored arrows denote direction and speed of elements of the gesture. Circle and Triangle shaped gestures exist in four, respectively two, different temporal versions, creating, together with similar shapes like the rectangle, a very challenging dataset for recognition and verification tasks. In Fig. 4 examples of all four circle shaped gestures are presented.

Fig. 3. Gesture set used for KinectDB (blue arrows indicating execution speed). (Color figure online)

The six participants were asked to perform each gesture twenty times, standing two meters in front of the Kinect device. The beginning and end of each gesture was manually determined by the experiment supervisor. The features included in the recording are the x, y and z coordinates of the right hand Kinect node and the timestamp of each recorded frame. Frames are recorded with a frequency of 30 Hz.

4.2 Experiment Protocol

For a better understanding of the experiment results, this section describes the details and circumstances of how they were obtained and what they consist of.

Performance Assessment. The evaluation of our CHnMM trajectory verification approach is the main goal of this work, utilizing real world authentication data. Furthermore, the possible application to three dimensional movement trajectory data is shown by using the KinectDB data set. For the assessment of the quality of authentication systems two important measures exist: the False Rejection Rate (FRR) of genuine trajectories and the False Acceptance Rate (FAR) of forgery trajectories which are usually used [9,12,15]. Commonly, a certain threshold value is employed in authentication systems, deciding whether a certain input fits the template. Depending on that threshold either a better FAR or a better FRR of the systems is favoured, rendering both values inversely related. Therefore, the so called Equal Error Rate (EER) where FAR equals FRR is provided as a single quantity to specify the quality of an authentication system, although it hides a lot of its actual behaviour.

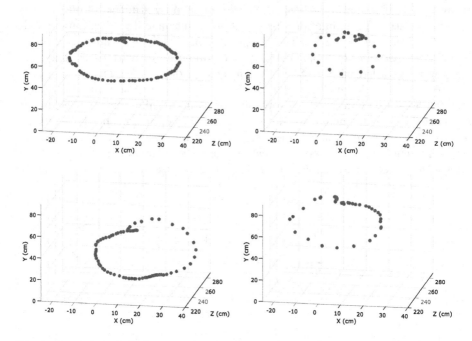

Fig. 4. Examples of all four circular shaped trajectories (Circle, Circle (fast), Circle (fast-slow), Circle (slow-fast)).

Input Data. In the previous section the data sets used for the experiments were introduced and explained, yielding four different corpora of interest: MCYT Signatures, Doodles, PseudoSignatures and KinectDB. Since they share a lot of similarities, one general experiment protocol is applied on them. Except for one, all corpora contain several genuine examples of a certain user trajectory, i.e. signature, doodle, pseudo signature or gesture, and also several forgeries of these user trajectories for each user. The exception is KinectDB that only has genuine examples for every user. In order to create a similar scale for the coordinates from all the different corpora, normalization has been applied, creating real valued coordinate ranges from 0 to 1.

For the experiments the trajectory data of each corpora needs to be divided into a training, a genuine and a forgery test set. Obviously, the training set is utilized to create and initialize the verification system, whereas both test sets are needed to determine the verification performance, specifically FAR and FRR. Inspired by the procedure in [12], two approaches for creating the test sets were used in the experiments: *random* and *skilled*. While for both the training and genuine set is similarly created by taking a specified number of genuine training examples from each user and using the remaining examples for the genuine set, the forgery set is created differently to test different qualities of forgeries. For the *random* forgery test set the first genuine example of every other user is taken and the performance results will help to understand the robustness of the verification system against random input. The *skilled* forgery set consists of all available forgery examples for the user and the results will reveal the applicability of the verification system in real world situations. For the KinectDB corpus that does not contain skilled forgery examples, a *skilled* forgery set is created by using the execution of the same gesture of other users. Those are assumed to be like possible real skilled forgery attempts.

Parameter Variation. The CHnMM authentication system that is described in this work has several parameters that influence the authentication behaviour. In order to determine acceptable parameter sets and to evaluate the influences of certain parameters, parameter variation has been utilized, hence, the system is tested with a lot of different parameter combinations. The tested parameter ranges are based on experience from previous work [4] and are as follows:

- **nRegions:** 10–20, step size 5,
- **minRadius:** 0.01–0.19, step size 0.02,
- **toleranceFactor:** 1.1–2.1, step size 0.2,
- **distributionType:** uniform and normal.

As a result, there are 360 different parameter sets that are used to evaluate the CHnMM authentication system. Additionally, to test the influence of a different number of training examples the experiments have been conducted with either five or ten training examples per user. Consequently, for each database corpus (MCYT, Doodles, PseudoSignatures, KinectDB) and forgery data type (*random* or *skilled*) $360 * 2$ FAR-FRR pairs are calculated. Plotting these results in a FAR-FRR point diagram helps to interpret the results. This diagram must not be confused with the so called Receiver Operating Characteristic (ROC) curve although it can seem very similar. The ROC curve is commonly used to visualize the behaviour of a verification system but in this work there is currently no single threshold parameter implemented.

4.3   Results

Result Overview. In Fig. 5 a FAR-FRR point diagram for every database corpus is presented to give an overview of the outcome of the conducted experiments. The diagrams focus on the most interesting result space with FAR and FRR below 50%, hence, several FAR-FRR points are not shown. The results visually resemble a typical ROC curve if only the Pareto frontier of most optimal results is considered. This general behaviour is as expected, because trying to reduce the FRR causes higher FAR and vice versa. Data points beyond the Pareto frontier are the result of bad parameter sets. Accordingly to the expectations, the performance difference between *random* (circles) and *skilled* (crosses) forgeries is obvious and similar across all corpora, with the FAR being very close to 0 for the *random* ones.

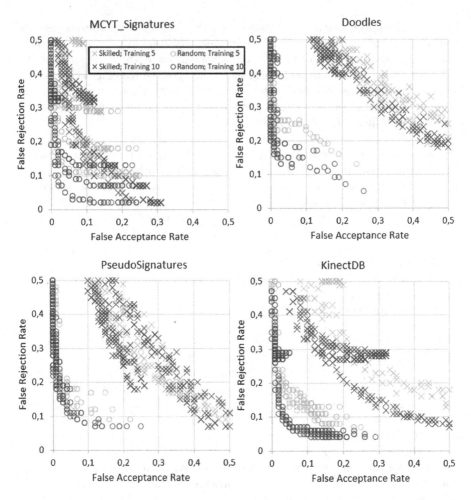

Fig. 5. FAR-FRR plots for all authentication experiment results distinguished by forgery type and training size. (Color figure online)

The best overall authentication performance of all corpora was achieved with the MCYT signatures, where the FAR and FRR values are generally the lowest. As a result, the difference between *random* and *skilled* is rather small compared to doodles, pseudo-signatures and KinectDB, where this difference is more remarkable. This could be explained by the fact that signatures written with a pen are performed more consistently, due to them being a common and known movement for the user. The pseudo-signature results are probably slightly better than the doodle results for the same reason, as pseudo-signatures are not performed as consistent as the signatures. With the KinectDB results generally being a little better than the pseudo-signature ones the applicability of the CHnMM verification approach to three dimensional data is proven.

Another unsurprising observation is that increasing the number of training examples from five (yellow) to ten (blue) generally improves the performances on all data sets, indicating that the developed system works as expected.

In Table 1 the achieved EER for each corpus and forgery type are presented. Be aware that these EER values describe the most balanced (FAR equals FRR) result that occurred in the parameter variation. The achieved EER values do not recommend to use the system in practice, especially due to the quite high percentages for the *random* forgeries that seemingly suggest that not even random input can be distinguished well. However, the plots prove that in all corpora the FAR values are very low for random inputs until the parameter sets become more tolerant. Hence, in order to better understand the values they have to be compared to other methods.

Table 1. Achieved EER for every database.

	MCYT	Doodles	PseudoSignatures	KinectDB
Random	4%	12%	8%	6%
Skilled	11%	29%	21%	20%

Martinez-Diaz et al. [12] created the DooDB and therefore also provided several benchmark values for the Doodle and Pseudo-signature corpora using a DTW verification approach. Fortunately, one basic DTW approach only utilized the two dimensional coordinates, or the first or second derivative respectively. This circumstance allows for a fairer comparison, as these features are not application specific like our approach that is not specialized on certain types of trajectories. The results are based on experiments with five training examples. Using skilled forgeries EER values between 26.7%–36.4% for doodles and between 19.8%–34.5% for Pseudo-signatures were achieved. For random forgeries the EER are between 2.7%–7.6% for doodles and between 1.6%–5.0% for Pseudo-signatures.

In the work by Ortega-Garcia et al. [15] an HMM verification approach was applied to subsets of the MCYT database where models were trained using 10 training examples. Depending on the chosen subset, EER between 1% and 3% were achieved for skilled forgeries. While this value could not be achieved with our system we still think that the performance is very promising, especially considering that it is not specialized on signature trajectories and that there is still room for improvement by employing a threshold system. This idea is further discussed in Sect. 4.3. Moreover, the HMM system utilized other recorded data like azimuth, elevation and pressure of the pen in order to reach these results. In [7] it is stated that only using the x and y coordinates resulted in an EER of 10.37%.

Parameter Influences. The influence and behaviour of the CHnMM system parameters still very much resembles the observations made in previous work [4] where the system was applied to touch gesture classification tasks. The parameters *minRadius* and *toleranceFactor* influence the system behaviour the most as

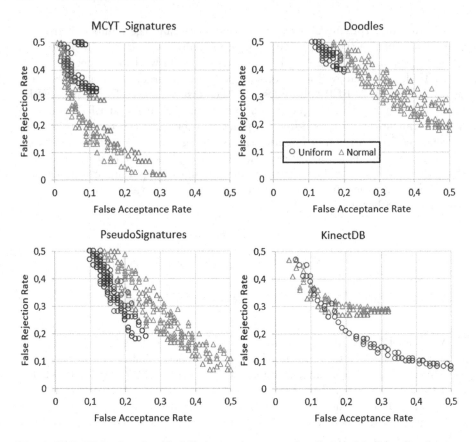

Fig. 6. FAR-FRR plots for all skilled experiment results distinguished by distribution type.

increasing their values generally create more tolerant verification systems that is more accepting and thus leads to lower FRR and higher FAR. Interestingly, parameter *nRegions* does not have a significant influence for certain parameter combinations especially those that lead to practically useless results with FAR greater than 50%, but a lower *nRegions* value can slightly improve the EER of the verification system for better parameter sets. This is due to the fact that a smaller number of regions in the model decreases the number of "hurdles" for a certain input and thereby the number of false rejections can be decreased while the chances of accepting an invalid input (FAR) only slightly increases.

In Fig. 6 the results of the experiments for skilled forgeries are plotted again but slightly different in order to analyse the influence of the distribution type of the transitions that are either *uniform* or *normal* in this work. The plots visualize that the uniform distribution generally seems to improve the FAR compared to the normal distribution while sacrificing on FRR. This is expected behaviour as the uniform distribution only covers a strict time interval while a normal distribution theoretically covers an infinite one. Hence, if the input does not fit

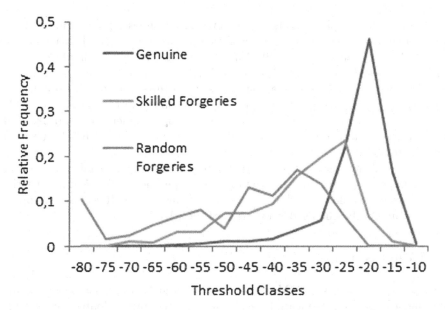

Fig. 7. Evaluation value distribution for a chosen parameter set with MCYT Signatures (logarithmised values) [5].

into the time interval at one point in the trajectory model the input is determined invalid. With the normal distribution such an early rule out by time cannot occur. The uniform distribution seems to perform better for the Pseudo-signatures which leads to think that the temporal behaviour is quite decisive in this data set. The same trend occurs in the Doodle database but an EER is never reached. For the MCYT signatures the normal distribution seems to be the better choice which probably is due to an unsuitable time tolerance for this data set.

Employing a Threshold Value. Currently, the implemented system does not employ the usual threshold concept as it is currently not decided how a threshold is determined best for our system. To proof that there is further potential to improve the already promising system an additional experiment was conducted on the MCYT signature database. This time with the data of all available 100 users, 10 training examples and only with a specific parameter set. The chosen set (nRegions = 10, toleranceFactor = 1.7, minRadius = 0.05, distributionType = normal) achieved the best balanced result (FAR = 10%, FRR = 12%) for skilled forgeries in the previous experiments. In this additional experiment the evaluation values of each verification have been recorded.

The resulting FAR and FRR values essentially did not change and in Fig. 7 the histogram shows how often certain evaluation values occurred in relation to the number of made verifications whose evaluation value were not 0. Be aware that the logarithm was taken of the evaluation values in order to make the very little values more comprehensible and easier to visualise.

As expected, the plot reveals that the evaluation values of genuine inputs tend to be greater than those of skilled and random forgeries with close to 95% of them being between −40 and −10. While there is no perfect threshold value that separates the forgeries from the genuines, it is possible to achieve improvements especially for the FAR. For example, setting the threshold to −40 would keep the FRR at 12% (there is only a slight deterioration from 11.9% to 12.2%) while significantly improving the FAR to 6.5%. Choosing a higher threshold like −30 would further improve the FAR to 3% at the expense of the FRR that would increase to 16.7%.

These findings suggest that the implementation of a threshold value could further improve the results from the previous experiments. We assume that the plotted results would see a shift to the left, because the FAR seems to improve with a comparably smaller deterioration of the FRR.

5   Conclusions

Within this article a CHnMM approach for two and three dimensional online movement trajectory verification has been presented and evaluated on four different data sets: signatures, doodles, pseudo-signatures and Kinect gestures. The outcome of the conducted experiments was shown to be in competitive ranges compared to HMM and DTW methods that others already applied to these data sets, proving the applicability of the developed CHnMM for trajectory verification tasks. The EER values for random forgeries were not as competitive, but the discussed implementation of a threshold value provides significant potential for a general improvement of the results. In addition to the results already presented in [5], the applicability of the developed CHnMM verification approach to three dimensional data could be proven in this work by performing experiments with the KinectDB data set, which revealed promising results.

Moreover, the results revealed that the employed system parameters can be utilized to adjust the behaviour to the needs of a given scenario, either preferring better FAR or FRR. For example, the use of a uniform distribution generally improves the FAR values by limiting acceptance in the time domain. In future iterations of the systems a new parameter for time tolerance besides the already existing tolerance factor for the spatial domain could be implemented to even further tune the system for either more accurate timing and/or accurate trajectory shape discrimination.

In the future, the developed CHnMM creation method for movement trajectories might be generalized to work on any time series like DTW and HMM, but with a focus on also discriminating temporal dynamics. Advantages are the fast computations and the independence of regular time series with a fixed time step.

Acknowledgements. We would like to thank Suryakiran Maruvada and Patrick Haese for creating the KinectDB, as well as the Biometric Recognition Group - ATVS for providing us access to their databases.

References

1. Bosse, S., Krull, C., Horton, G.: Modeling of gestures with differing execution speeds: are hidden non-Markovian models applicable for gesture recognition. In: MAS: The International Conference on Modelling and Applied Simulation, pp. 189–194 (2011)
2. Buchholz, R.: conversive hidden non-Markovian models. Ph.D. thesis, Otto-von-Guericke-Universität Magdeburg (2012)
3. Dittmar, T., Krull, C., Horton, G.: Using conversive hidden non-Markovian models for multi-touch gesture recognition. In: The 12th International Conference on Modeling and Applied Simulation, September 2013
4. Dittmar, T., Krull, C., Horton, G.: An improved conversive hidden non-Markovian model-based touch gesture recognition system with automatic model creation. In: The 14th International Conference on Modeling and Applied Simulation, September 2015
5. Dittmar, T., Krull, C., Horton, G.: Evaluating a new conversive hidden non-Markovian model approach for online movement trajectory verification. In: The 6th International Conference on Pattern Recognition Applications and Methods (ICPRAM), February 2017
6. Faundez-Zanuy, M.: On-line signature recognition based on VQ-DTW. Pattern Recogn. **40**(3), 981–992 (2007)
7. Fierrez, J., Ortega-Garcia, J., Ramos, D., Gonzalez-Rodriguez, J.: HMM-based on-line signature verification: feature extraction and signature modeling. Pattern Recogn. Lett. **28**(16), 2325–2334 (2007)
8. Fink, G.A.: Markov Models for Pattern Recognition: From Theory to Applications. Springer, London (2014). https://doi.org/10.1007/978-1-4471-6308-4
9. Kholmatov, A., Yanikoglu, B.: SUSIG: an on-line signature database, associated protocols and benchmark results. Pattern Anal. Appl. **12**(3), 227–236 (2009)
10. Kholmatov, A., Yanikoglu, B.: Identity authentication using improved online signature verification method. Pattern Recogn. Lett. **26**(15), 2400–2408 (2005)
11. Krull, C., Horton, G.: Hidden non-Markovian models: formalization and solution approaches. In: Proceedings of 6th Vienna International Conference on Mathematical Modelling (2009)
12. Martinez-Diaz, M., Fierrez, J., Galbally, J.: The DooDB graphical password database: data analysis and benchmark results. IEEE Access **1**, 596–605 (2013)
13. Martinez-Diaz, M., Fierrez, J., Krish, R.P., Galbally, J.: Mobile signature verification: feature robustness and performance comparison. IET Biom. **3**(4), 267–277 (2014)
14. Muramatsu, D., Matsumoto, T.: An HMM online signature verifier incorporating signature trajectories. In: Proceedings of the Seventh International Conference on Document Analysis and Recognition, pp. 438–442. IEEE (2003)
15. Ortega-Garcia, J., Fierrez-Aguilar, J., Simon, D., Gonzalez, J., Faundez-Zanuy, M., Espinosa, V., Satue, A., Hernaez, I., Igarza, J.J., Vivaracho, C., et al.: MCYT baseline corpus: a bimodal biometric database. IEE Proc.-Vis. Image Sig. Process. **150**(6), 395–401 (2003)
16. Rabiner, L., Juang, B.: An introduction to hidden Markov models. IEEE ASSP Mag. **3**(1), 4–16 (1986)
17. Tian, J., Qu, C., Xu, W., Wang, S.: KinWrite: handwriting-based authentication using kinect. In: NDSS (2013)

Prediction of User Interest by Predicting Product Text Reviews

Esteban García-Cuesta[1](✉), Daniel Gómez-Vergel[1], Luis Gracia-Expósito[1], José Manuel López-López[1], and María Vela-Pérez[2]

[1] Data Science Laboartory, School of Arquitecture, Engineering and Design, Universidad Europea de Madrid, Calle Tajo S/N Villaviciosa de Odón, 28670 Madrid, Spain
{esteban.garcia,daniel.gomez,luis.gracia, josemanuel.lopez}@universidadeuropea.es
[2] Instituto de Matemáticas Interdisciplinar, Departamento de Estadística e Investigación Operativa II, Facultad de Ciencias Económicas y Empresariales, Universidad Complutense de Madrid, 28040 Madrid, Spain
maria.vela@ucm.es

Abstract. Most item shopping websites currently provide social network services (SNS) to collect their users' opinions on items available for purchasing. This information is often used to reduce information overload and improve both the efficiency of the marketing process and user's experience by means of user-modeling and hyper-personalization of contents. Whereas a variety of recommendation systems focus almost exclusively on ranking the items, we intend to extend this basic approach by predicting the sets of words that users would use should they express their opinions and interests on items not yet reviewed. To this end, we pay careful attention to the internal consistency of our model by relying on well-known facts of linguistic analysis, collaborative filtering techniques and matrix factorization methods. Still at an early stage of development, we discuss some encouraging results and open challenges of this new approach.

Keywords: User opinion · Recommendation systems · Prediction
Hyper personalization · User modeling · Big data

1 Introduction

With the advent of the Internet and its social websites, e-commerce web sites have enabled users to share their opinions with other customers, mostly by allowing them to submit scores and opinion reviews that may help potential purchasers pick the most suitable items. This is done by recommending new products they may be interested in [20] or by identifying other users of similar taste in order to make recommendations based on those similarities [26]. These actions lead to the

© Springer International Publishing AG, part of Springer Nature 2018
M. De Marsico et al. (Eds.): ICPRAM 2017, LNCS 10857, pp. 132–146, 2018.
https://doi.org/10.1007/978-3-319-93647-5_8

hyper-personalization of the website, better marketing strategies, and improved users' experience.

Recommender systems may follow different strategies to model users, including defining the set of features that best describe them – such as tags, keywords, text comments, and likes/dislikes [17] – or, in more sophisticated systems, by automatically unveiling the latent structure that captures the users' interests based on how they evaluate and review products [3]. This latter case typically uses plain-text reviews and/or numerical scores, together with machine learning algorithms, to predict the scores that users will give to items still unreviewed [13].

The interested reader will find a vast literature devoted to these topics. In [21], for instance, authors present a hidden factor model to understand why any two users may agree when reviewing a movie, yet disagree when reviewing another: The fact that users may have similar preferences towards one genre, but opposite preferences for another turns out to be of primary importance in this context. Also, [21] proposes the use of the aforementioned latent factors to achieve a better understanding of the rating dimensions to be connected with the intrinsic features of users and their tastes. Other authors [8,14,22] use text reviews to better understand user *sentiments*, hence improving the user modeling process to generate ratings. The rapid proliferation of social media has gone hand in hand with the development of sentiment analysis techniques [15], which have been successfully applied in a wide range of fields, from social networks [12] to movies [5].

In this article, we intend to predict the representative sets of words that users will choose to express their opinions on non-previously reviewed items, naturally extending previous models that use latent spaces to predict or support ratings [27,28]. Indeed, predicting the future opinion (or text reviews) on unreviewed products remains, to the best of our knowledge, an open quest that deserves further attention [31].[1] Our approach – that assumes the existence of a latent space that accurately represents the users' interests and tastes [21] – is based on a two-step process:

1. Setting up the opinion dictionary, not too large as to impede numerical computations, but rich enough as to characterize the user's opinions. Take the words 'expensive' and 'good_quality' for example,[2] the former being a purely subjective term which expresses a negative opinion about a product, the latter expressing a positive opinion instead. We would like these terms to be part of the dictionary since they convey relevant information on the user's opinion.
2. Predicting the set of words users would choose should they have the opportunity to review an item, based on the hidden dimensions that represent their tastes.

---

[1] From now on, we will use the *opinion* and *review* terms without distinction whenever we refer to the text (or a representative part of it) that a user writes about a product.
[2] We work with opinion-words provided by a natural language analysis tool. Terms may be compositions of several words.

1.1 Recommendation Systems Review

A rich collection of algorithms and recommender systems has been developed over the last two decades. The wide range of domains and applications shows that there is not a one-size-fits-all solution to the recommendation problem and that a careful analysis of prospective users and their goals is necessary to achieve good results. Most existing recommendation systems can be classified into two broad groups, namely: (i) content based recommender (CBR) and (ii) collaborative filtering (CF) systems.

The first approach recommends items based on some specific features/keywords that describe them and a user profile model that represents all the information available on users. In a basic problem setup, this includes the users' characteristics and interests on a set of items based on previous or current interactions with the system. The recommendations are then generated by comparing items with the users' profiles, thus predicting the products they will be interested in (see [17] for a detailed state of the art until 2010). Notice that one important limitation of this basic approach is the fact that it ignores users' opinions on different elements, taking only their characteristics and preferences into consideration. Indeed, most models rely on user profiles built *a priori* and used later on to predict the recommendations. This methodology, however, does not attempt to reflect the intrinsic likes and dislikes of users on different items, focusing on a more general description of their preferences instead.

Some benefits of this approach are its simplicity and ease of interpretation of the recommendations provided. Some recent works that make use of this type of models have tried to extend the range of applicability by combining this models with sentiment analysis [7] and also focusing on the dynamic aspects of users' profiles to make them evolves effectively over time [4].

The second approach – simply abbreviated as CF – has achieved the most successful results and focuses on users' behaviors as proposed by [29] rather than on the users' characteristics. This method uses the similarities among users to discover the latent model which best describes them and retrieves predicted rankings for specific items [18,19]. More specifically, CF is a technique that generates automatic predictions for a user by collecting taste information from other people [25]. The information domain for these systems consists of users who already expressed their preferences for various items, represented by ($user, item, rating$) triples. The rating is typically a natural number between zero and five or a Boolean (like/dislike) variable. The resulting associated rating matrix is usually subject to sparsity due to the existence of unrated items and the full evaluation process often requires the completion of two tasks: (i) predicting the unknown ratings and (ii) providing the best ranked list of n items for a given user [6].

This approach has found applications in areas such as social media recommendation [10] or recommending news articles [16]. As mentioned before, new lines of work focus on extracting interpretable textual labels from the latent factors of CF [21].

1.2   Linguistic Processing

Literature often divides opinion-words or *sentiments* into two categories: (*i*) rational and (*ii*) emotional sentiments:

- Rational sentiments, namely, "rational reasoning, tangible beliefs, and utilitarian attitudes" [1]. An example of this category is given by the sentence "This guitar is affordable", which does not involve emotions like happiness. In this case, the opinion-word (that is, the adjective "affordable") fully reveals the user's opinion.
- Emotional sentiments, described in [15] as "entities that go deep into people's psychological states of mind". For example, "I trust this camera". Here, the opinion-word is the "trust" verb, that clearly conveys the emotional state of the writer.

The main challenge in designing our automatic opinion prediction model is to define a vocabulary rich enough as to characterize the users' viewpoints, but not too large as to impede numerical computations. Concretely, given an item and the associated set of users who reviewed it, we carry out a sentiment analysis at the phrase level, thereby extracting the relevant opinion-words and creating a dictionary specifically for that item (see Fig. 1). This is in contrast with our previous approach to the model, where all items shared the same constant-length dictionary [9].

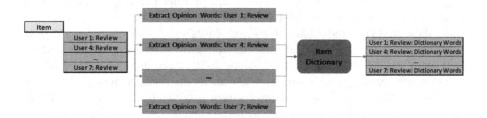

Fig. 1. Generation process of an item's dictionary.

The dictionary generation is achieved by using a solution graciously provided to us by Bitext,[3] a highly effective text analytics and linguistic technology [2] (see Table 2 as an example of retrieved words included in the dictionary). In a second step, we generate a feature vector for each user-item pair in our data set contain the frequency of occurrence of the opinion-words. In a second step, we generate a feature vector for each user-item pair in our dataset that contains the frequency of occurrence of the opinion-words.

It is worth pointing out that the set of terms used in this work follows the power law distribution known as Zipf's law [30, 32], that states that the frequency of occurrence of an opinion-word is inversely proportional to its ordering number (see Fig. 2).

---

[3] https://www.bitext.com.

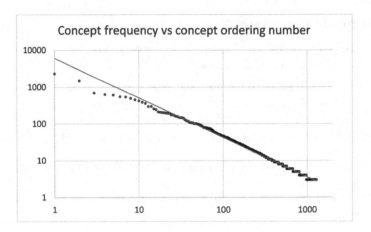

Fig. 2. Frequency of occurrence of the concepts considered in the Amazon's "Musical Instruments" dataset, as a function of their ordering number. The straight line shows a least squares fitting in perfect agreement with Zipf's law.

1.3 Contributions

Our main contribution is to propose and describe – for the very first time, to the best of the authors' knowledge – a model that combines the use of hidden dimensions – associated with users' tastes and product features – and a matrix factorization approach to predict the user's opinion on not reviewed items. The results show that the prediction of the set of words which best describes a review is possible and gives, at this early stage of development, an initial understanding of the main reasons why a user would like or dislike a product. This is important since this information can be used to complement the rating's value and provide extra information to the user whenever a new product is recommended. Thus, this approach could be used together with the current recommendation systems to provide further insight into the reasons why the product is recommended to a specific user, knowing that the very same product may be recommended to another user for completely different reasons.

The rest of the paper is organized as follows: Sect. 2 introduces the notation used throughout this article and reviews collaborative filtering approaches and the ALS matrix factorization process. Section 3 describes the experiments we conducted to test the implemented model. We show our results in Sect. 4 and discuss the model's strengths and weaknesses. Finally, Sect. 5 presents the conclusions and some insights into future work.

2 User Modeling Based on Opinions

In what follows, we introduce the terminology and notation used throughout this article. We then proceed to explain in full detail our new opinion prediction model based on tensor factorization.

2.1 Notation

A typical online shopping website with SNS capabilities provides, for the purposes of this article, N reviewers writing reviews on a set of M items. Generally, a given user will have scored and reviewed only a subset of these M items, thus making the website's database highly sparse.

Table 1. Notation.

Symbol	Description
K	Rank, number of latent dimensions
u	User, reviewer
N	Total number of users
i	Item, product
M	Total number of items
S	Set of (u, i) pairs of existing reviews
$\mathbf{t}_{ui}$	u-th user's review ('document') on i-th item
t_{ij}	j-th word describing the i-th item
f_{uij}	Frequency of occurrence of the j-th word in t_{ui}
D_i	Vocabulary size of the i-th product's dictionary
D	Sum $\sum_i^M D_i$ of dictionaries' lengths
$\mathbf{R}$	Input matrix in $\mathbb{R}^{N \times D}$

Let $S = \{(u, i) \mid u = 1, \ldots, N; i = 1, \ldots, M\}$ denote the set of user-item pairs for which written reviews do exist and let $\mathbf{t}_{ui}$ be their associated feature vectors. Our information domain consists then of triples of the form $(u, i, \mathbf{t}_{ui})$. In our model, different items are described by different sets of words, making them substantially vary in vocabulary size D_i. The $\mathbf{t}_{ui} \in \mathbb{R}^{D_i}$ vector is populated with the frequencies $f_{uij} \geq 0$ of occurrence of the $j = 1, \ldots, D_i$ words in the (u, i)-th review.

The website's 2-dimensional input matrix $\mathbf{R}$ is set up by concatenating the $N \times D_i$ sparse matrices $\mathbf{R}_i$ containing the reviews for the $i = 1, \ldots, M$ products, that is,

$$\mathbf{R} = [\mathbf{R}_0 \ \mathbf{R}_1 \ \cdots \ \mathbf{R}_M] \in \mathbb{R}^{N \times D},$$

where $D = \sum_i D_i$ denotes the sum of the vocabulary sizes for the M products. The $\mathbf{R}$ matrix represents a high dimensional space where users' opinions – either positive or negative – are latent and can be represented by a subset of new features in a lower dimensional space. Table 1 summarizes the notation; see also Fig. 3 for further clarification.

	Item 1			Item 2						Item M		
	t_{11}	t_{12}		t_{21}	t_{22}	t_{23}	t_{24}			t_{M1}	t_{M2}	t_{M3}
User 1	?	?	User 1	f_{121}	f_{122}	f_{123}	f_{124}		User 1	f_{1M1}	f_{1M2}	f_{1M3}
User 2	f_{211}	f_{212}	User 2	?	?	?	?	...	User 2	?	?	?
User 3	f_{311}	f_{312}	User 3	f_{321}	f_{322}	f_{323}	f_{324}		User 3	f_{3M1}	f_{3M2}	f_{3M3}
...			...						...			
User N	f_{N11}	f_{N12}	User N	f_{N21}	f_{N22}	f_{N23}	f_{N24}		User N	?	?	?

Fig. 3. Input matrix $\mathbf{R}$ is obtained by placing the tables of frequencies for different items side by side. Reviews whose entries are labeled with a question mark '?' are removed from the matrix for testing or validation. Notice that items have distinct sets of words of variable length. In this figure, $D_1 = 2$, $D_2 = 4$, and $D_M = 3$.

2.2 Predicting the Opinion Using Alternating Least Squares (ALS)

Our model follows a collaborative filtering approach and uses the Alternating Least Squares (ALS) procedure to predict users' opinions (that is, $\mathbf{t}_{ui}$ vectors) not included in S. Our objective is, therefore, to generate automatic predictions for a given user by collecting taste information from other reviewers – see [6] for more information.

Specifically, we subject the input matrix $\mathbf{R}$ to an ALS factorization [13, 24] of the form $\mathbf{R} \approx \mathbf{P}\mathbf{Q}^{\mathrm{T}}$ in order to estimate the missing reviews. Here, $\mathbf{P} \in \mathbb{R}^{N \times K}$ and $\mathbf{Q} \in \mathbb{R}^{D \times K}$, where $K \in \mathbb{N}$ is the number of latent factors or features [3] – in our model, a predefined constant typically close to ten. Any frequency f_{uij} can then be approximated by the usual scalar product $\hat{f}_{uij} = \mathbf{p}_u^{\mathrm{T}}\mathbf{q}_{ij}$, with $\mathbf{p}_u \in \mathbb{R}^{K \times 1}$ the u-th row of $\mathbf{P}$ and $\mathbf{q}_{ij} \in \mathbb{R}^{K \times 1}$ the $(\sum_1^i D_k + j)$-th row of $\mathbf{Q}$. The procedure minimizes the quadratic loss function

$$\langle P^*, Q^* \rangle = \arg\min_{\mathbf{P},\mathbf{Q}} \sum_{(u,i)\in S} \left(\lambda \mathbf{p}_u^{\mathrm{T}}\mathbf{p}_u + \sum_{j=1}^{D_i} \left(\epsilon_{uij}^2 + \lambda \mathbf{q}_{ij}^{\mathrm{T}}\mathbf{q}_{ij} \right) \right),$$

where $\epsilon_{uij} = f_{uij} - \hat{f}_{uij}$. Here, λ denotes the regularization parameter in the ALS method that balances the training error and the size of the solution.

As explained in detail in [24], the ALS technique alternates between $\mathbf{P}$-steps – where $\mathbf{Q}$ is fixed and $\mathbf{P}$ is recomputed by solving a least-square problem – and $\mathbf{Q}$-setps – where the previous order of computation is reversed. More specifically, for a $\mathbf{P}$-step, let $\mathbf{Q}_u \in \mathbb{R}^{D_u \times K}$ be the restriction of $\mathbf{Q}$ to the items reviewed by the u-th user. Here, D_u denotes the sum of the dictionaries' lengths associated with those items, that is,

$$D_u = \sum_{i:(u,i)\in S} D_i.$$

The u-th column of $\mathbf{P}$, $\mathbf{p}_u \in \mathbb{R}^{K \times 1}$, is then recomputed as

$$\mathbf{p}_u = \left(\lambda D_u \mathbf{I}_K + \mathbf{Q}_u^{\mathrm{T}}\mathbf{Q}_u \right)^{-1} \mathbf{Q}_u^{\mathrm{T}}\mathbf{t}_u.$$

Here, $\mathbf{I}_K$ denotes the $K \times K$ identity matrix and $\mathbf{t}_u \in \mathbb{R}^{D_u \times 1}$ is the concatenation of the $\mathbf{t}_{ui}$ feature vectors. A $\mathbf{Q}$-step operates in a similar manner.

3   Experiments

For motivational purposes, we test our model using the musical instruments Amazon dataset, which contains user-product-rating-review quads for a total of $10,261$ reviews for 1429 users and 900 products [23][4]. We chose this dataset due to its ease of interpretability and reasonable size.[5]

At a first step, we process the reviews using the Bitext natural language processor making all the basic tokenization, lemmatization, PoS [2], and concept identification tasks straightforward. This enables us to syntactically analyze the texts in an efficient manner in order to extract the simple (*e.g.*, 'cheap') and compound (*e.g.*, 'highly_expensive') opinion words to be part of the dictionary.

At this stage, we keep track of the (usually different) sets of words used by each customer in their product reviews, along with their frequencies of occurrence. The dictionary associated with a specific item is obtained by taking the union of the individual users' lists of words for that item. To retain the most relevant words only and keep the complexity of the problem manageable, we opt to discard all words with frequencies of occurrence below a threshold frequency $f_{\min} = 3$. For instance, a total of 1285 opinion words remain after this selection process for the "Musical Instrument" dataset (see Table 2).

Table 2. Some concepts with a high frequency of occurrence.

'great', 'good', 'nice', 'easy', 'love', 'best', 'perfect',
'cheap', 'fine', 'solid', 'worth', 'excellent', 'problem',
'sturdy', 'recommend', 'long', 'awesome', 'very_good',
'fit', 'ok', 'wrong', 'simple', 'amazing', 'noise',
'would_recommend', 'happy', 'very_nice', 'favorite',
'very_happy', 'decent', 'clean', 'bad', 'really_like'
'inexpensive', 'durable', 'fantastic', 'strong', 'adjustable'.

Next, we randomly remove 10% of the word lists – that is, $\mathbf{t}_{ui}$ reviews – from the 2-dimensional $\mathbf{R}$ matrix to select the optimal values of the model's parameters – namely, λ, K, and the number of ALS iterations, giving raise to a train set $\mathbf{R}_{train}$ and a test set $\mathbf{R}_{test}$. The way to achieve this is straightforward:

---

[4] http://jmcauley.ucsd.edu/data/amazon/.

[5] We use 20 executors with 8 cores and 16 GB RAM on a Hadoop cluster with a total of 695 GB RAM, 336 cores, and 2 TB HDFS. Our implemented algorithms are easily scalable, so any RAM limitation might be solved using a cluster with a sufficiently large number of nodes.

We pick 10% of all user-item coordinates $(u, i) \in S$ at random – a total of 1024 pairs – and replace their corresponding word lists with empty vectors. The resulting sparse matrix $\mathbf{R}_{train}$ is finally subjected to ALS factorization in order to reconstruct the removed vectors and compare them with the originals. We use mean square errors (MSE) to determine the quality of the predictions.

The degree of predictability of our model is assessed by means of a Jaccard index that measures the similarity between the original and predicted reviews in the following sense: If the frequencies in the predicted and the original (u, i, j)-entries of the input matrix are both positive or both zero, then their Jaccard distance is set to zero; otherwise it is one.

All our codes are implemented in Python 3.5 using the collaborative filtering RDD-based Apache Spark implementation of the ALS algorithm [6], which is well known for its robustness and efficiency. This implementation, in turn, makes use of the MLlib library[7].

4   Results

When deciding on a choice for the K latent factors, we must reach a compromise between the model's error and the running time of the ALS method, which is proportional to K^3 [11]. For the database used in this article, $K = 20$ seems to be the optimal choice. In what follows, all graphs use $\lambda = 0.1$ as the optimal value for the regularization parameter – see Fig. 4 for more details.

Figure 5 displays several MSE curves relative to the number of ALS iterations for different values of K. As expected, graphs are strictly decreasing for low numbers of iterations and become nearly horizontal for sufficiently large values of them. Similar MSE graphs for the test results – see Fig. 6 – attain their absolute minima between two and three iterations. In this case, the small size of the used dataset and the efficiency of the ALS algorithm explain the rapid convergence of the model. These minima are nonetheless expected to shift toward greater numbers of iterations for larger databases.

We use the Jaccard distance to evaluate whether a word appears or not in the prediction. In matrix language, if the (u, i, j)-th entries of the test $\mathbf{R}_{test}$ and training $\mathbf{R}_{train}$ sets are both positive or both zero, then the Jaccard is zero; otherwise, it is 1. Figure 7 shows the frequency of reviews with a given Jaccard index (the reader is referred to Sect. 3 to check its definition). The histogram displays the expected hyperbolic decay for small index values, but we observe some irregularities for larger values, particularly at $1/2$ – something that may be explained by the very definition of Jaccard distance and the finite length of the dictionaries – and 1 – a behavior probably dominated by the interchange of synonyms in the reviews that affect the overall accuracy of the model.

In a complementary approach, it is possible to introduce a discriminative threshold α as an independent super-parameter of the model. When the predicted value of a word surpasses this threshold, we interpret that the model

---

[6] It is part of the MLlib Apache's Library.

[7] Apache Spark's scalable machine learning library.

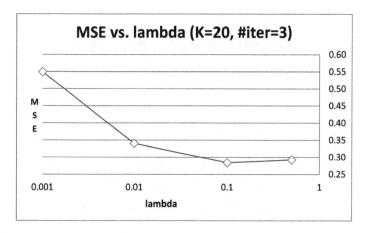

Fig. 4. MSE for the test set versus the regularization parameter λ. $\lambda = 0.1$ is consistently found to be its optimal value for $K = 20$ and three iterations.

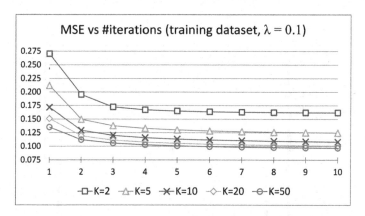

Fig. 5. MSE for the training set versus the number of ALS iterations. The shown latent factors are $K = 2, 5, 10, 20, 50$, and parameter $\lambda = 0.1$. Notice how the vertical distances between adjacent curves decrease drastically for increasing latent factors, making $K = 20$ an optimal choice for the number of latent factor.

predicts the appearance of that word. The predicted condition can then be compared with the true condition of that word. The results are summarized in Table 3. Changing the super-parameter α produces a trading between *precision* (positive predictive value) and *recall* (true positive rate). In this work, we consider that both recall and precision are equally important to assess the effectiveness of the model as a classifier system. Thus, we choose the α value that produces the largest F_1 measure, which is the harmonic mean of precision and recall. For comparative purposes, the best F_1 measure obtained from the training dataset is 0.726.

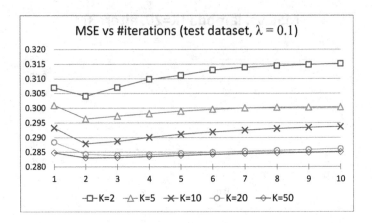

Fig. 6. MSE for the test set versus the number of ALS iterations. The shown latent factors are $K = 2, 5, 10, 20, 50$, and parameter $\lambda = 0.1$.

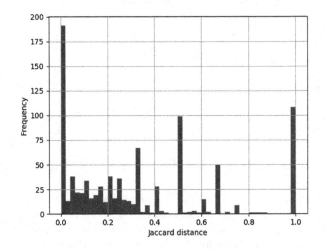

Fig. 7. Frequency of reviews with a given Jaccard index for $K = 20, \lambda = 0.1$ and three iterations.

Table 3. Precision, recall, accuracy, and F-measure of the model with $K = 20, \lambda = 0.1$ and 3 iterations.

$\alpha = 0.110$		Predicted		Precision $= 0.247$
		Positive	negative	Recall $= 0.433$
True	Positive	544	711	Accuracy $= 0.739$
	Negative	1661	6183	$F_1 = 0.315$

We want to highlight however that, due to the large vocabulary size and sparsity of the data, many reviews are predicted as zero vectors, even though they contain nonzero frequencies in the training set. This makes it advisable to use a variant of the ALS method specifically optimized for low-rank matrices, a problem that we attempt to address in the future.

5 Conclusions

The fundamental hypotheses behind our model is that it is possible to predict a user's review on a previously unreviewed product by means of a CF based model. Our proposed method makes use of state-of-the-art NLP tools and implements a new ALS-based model to unveil the latent dimensions that best represent the users' expressiveness. Ideally, the different reasons that lead to a user's opinion can be captured by those latent factors, and hence, they can be predicted through a direct comparison with other users of similar taste. An important feature that distinguishes this model from our previous attempts, is the introduction of a distinct pre-built opinion dictionary for each item that contains its most representative opinion-words.

The results show that the model, although still at a preliminary stage of development, is able to deduce the latent dimensions and to provide predictions meaningful enough as to gain useful insights into the potential opinions of users on new products. The model certainly calls for further improvements, however. For instance, when it is used as a classifier for predicting the occurrence or not of a word, the F_1-score takes a value around 30% for the test set in our experiments (see Table 3). Moreover, the overall accuracy, which takes into account both true positive and true negative cases, is approximately 74%. This shows that the model has a tendency to overlearn the zero values of the matrix, associated with the words not used in the reviews. Notice that in order to predict the users' interests on a product, we must be able to predict both the positive and the negative cases. Otherwise, linking a word to a review when it is not actually relevant will hinder the prediction of the user's true opinion.

There are some lines of work already in progress that we hope will invite future improvements:

1. There are suggestive indications that columns in $\mathbf{R}_{test}$ corresponding to semantically equivalent words in the same item's dictionary (synonyms) may have been projected into a small neighborhood of the same latent space, thus making them randomly interchangeable in the predicted reviews. We intend to subsume such equivalent words in a single concept in order to avoid these undesired exchanges.
2. We expect more accurate predictions for larger datasets. In this context, the model may benefit from a larger number of reviews (at a fixed dictionary length) and further adjustment of the items' dictionaries, rethinking how the most relevant opinion-words must be selected from a semantical point of view. Indeed, concepts of low descriptive value should better be avoided if this approach is really to be used to provide predictions in recommendation systems.

Finally, we mention the scalability of this new approach – implemented using a Hadoop based cluster and its distributed computational and storage resources. We intend to conduct further and more exhaustive analysis in larger datasets by enlarging these capabilities, in the hope that we will obtain better statistics this way. It is our believe that a deeper analysis of the latent factors and their categorization will also allow a better understanding of the conceptual parts of a language involved in the users' opinions.

Acknowledgements. The authors want to thank Bitext (http://bitext.com) for providing NLP services for research. They also acknowledge the support of the Universidad Europea de Madrid through the E-Modelo research project. Special thanks to Hugo Ferrando for developing a significant part of the code used for experimentation, and to Javier García-Blas for their insightful comments. This work has been partially supported by the Spanish7 Ministry of Economy and Competitiveness through the MTM2014-57158-R project.

References

1. Almashraee, M., Monett Díaz, D., Paschke, A.: Emotion level sentiment analysis: the affective opinion evaluation. In: Joint Proceedings of the 2th Workshop on Emotions, Modality, Sentiment Analysis and the Semantic Web and the 1st International Workshop on Extraction and Processing of Rich Semantics from Medical Texts Co-located with ESWC 2016, Heraklion, Greece, 29 May (2016)
2. Benjamins, R., Cadenas, D., Alonso, P., Valderrabanos, A., Gomez, J.: The voice of the customer for digital telcos. In: Proceedings of the Industry Track at the International Semantic Web Conference (2014)
3. Bennet, J., Lanning, S.: The netflix prize. In: KDD Cup and Workshop (2007)
4. Cami Bagher, R., Hassanpour, H., Mashayekhi, H.: User trends modeling for a content-based recommender system. Expert Syst. Appl. **87**, 209–219 (2017). ISSN 0957–4174
5. Diao, Q., Qiu, M., Wu, C.Y., Smola, A.J., Jiang, J., Wang, C.G: Jointly modeling aspects, ratings and sentiments for movie recommendation (JMARS). In: KDD 2014, pp. 193–202. https://doi.org/10.1145/2623330.2623758
6. Ekstrand, M.D., Riedl, J.T., Konstan, J.A.: Collaborative filtering recommender systems. Found. Trends Hum.-Comput. Interact. **4**, 81–173 (2012)
7. Feltoni Gurini, D., Gasparetti, F., Micarelli, A., Sansonetti, G.: A sentiment-based approach to Twitter user recommendation. In: Proceedings of the International ACM Recommendation Systems Conference, RecSys (2013)
8. Ganu, G., Elhadad, N., Marian, A.: Beyond the stars: improving rating predictions using review text content. In: WebDB (2009)
9. García-Cuesta, E., Gómez-Vergel, D., Gracia-Expósito, L., Vela-Pérez, M.: Prediction of user opinion for products - a bag of words and collaborative filtering based approach. In: Proceedings of the 6th International Conference on Pattern Recognition Applications and Methods ICPRAM 2017, pp. 233–238. Science and Technology Publications (2017)
10. Guy, I., Zwerdling, N., Ronen, I., Carmel, D., Uziel, E.: Social media recommendation based on people and tags. In: Proceedings of the 33rd International ACM SIGIR Conference on Research and Development in Information Retrieval (SIGIR 2010), pp. 194–201. ACM, New York (2010)

11. Hu, Y., Koren, Y., Volinsky, C.: Collaborative filtering for implicit feedback datasets. In: 8th IEEE International Conference on Data Mining, IDCM 2008, Pisa, Italy, pp. 263–272 (2008)
12. Hu, X., Tang, J., Gao, H., Liu, H.: Unsupervised sentiment analysis with emotional signals. In: WWW 2013, pp. 607–618. https://doi.org/10.1145/2488388.2488442
13. Koren, Y., Bell, R., Volinsky, C.: Matrix factorization techniques for recommender systems. J. Comput. **42**, 30–37 (2009)
14. Ling, G., Lyu, M.R., King, I.: Ratings meet reviews, a combined approach to recommend. In: RecSys, vol. 14, pp. 105–112 (2014) https://doi.org/10.1145/2645710.2645728
15. B. Liu, Sentiment Analysis: Mining Opinions, Sentiments, and Emotions. Cambridge University Press, Cambridge (2015)
16. Shan, L., Dong, Y., Chai, J.: Research of personalized news recommendation system based on hybrid collaborative filtering algorithm. In: 2016 2nd IEEE International Conference on Computer and Communications (ICCC). IEEE (2016)
17. Lops, P., de Gemmis, M., Semeraro, G.: Content-based recommender systems: state of the art and trends. In: Ricci, F., Rokach, L., Shapira, B., Kantor, P.B. (eds.) Recommender Systems Handbook, pp. 73–105. Springer, Boston (2011). https://doi.org/10.1007/978-0-387-85820-3_3
18. Luo, X., Zhou, M., Shang, M., Li, S., Xia, Y.: A novel approach to extracting non-negative latent factors from non-negative big sparse matrices. IEEE Access **4**, 2649–2655 (2016)
19. Luo, X., Zhou, M., Shang, M., Li, S., You, Z., Xia, Y., Zhu, Q.: A nonnegative latent factor model for large-scale sparse matrices in recommender systems via alternating direction method. IEEE Trans. Neural Netw. Learn. Syst. **27**(3), 579–592 (2015). https://doi.org/10.1109/TNNLS.2015.2415257
20. McAuley, J., Leskovec, J.: From amateurs to connoisseurs: modeling the evolution of user expertise through online reviews. In: WWW (2013)
21. McAuley, J., Leskovec, J.: Hidden factors and hidden topics: understanding rating dimension with review text. In: Proceedings of the 7th ACM Conference on Recommender Systems, RecSys (2013)
22. McAuley, J., Leskovec, J., Jurafsky, D.: Learning attitudes and attributes from multi-aspect reviews. In: ICDM (2012)
23. McAuley, J., Pandey, R., Leskovec, J.: Inferring networks of substitutable and complementary products. In: Knowledge Discovery and Data Mining (2015)
24. PiFlászy, I., Zibriczky, D., Tikk, D.: Fast ALS-based matrix factorization for explicit and implicit feedback datasets. In: Proceedings of the 4th ACM Conference on Recommender Systems, pp. 71–78. ACM (2010)
25. Schafer, J.B., Frankowski, D., Herlocker, J., Sen, S.: Collaborative filtering recommender systems. In: Brusilovsky, P., Kobsa, A., Nejdl, W. (eds.) The Adaptive Web. LNCS, vol. 4321, pp. 291–324. Springer, Heidelberg (2007). https://doi.org/10.1007/978-3-540-72079-9_9
26. Sharma, A., Cosley, D.: Do social explanations work? Studying and modeling the effects of social explanations in recommender systems. In: WWW (2013)
27. Titov, I., McDonald, R.: A joint model of text and aspect ratings for sentiment summarization. In: ACL (2008)
28. Wang, H., Lu, Y., Zhai, C.: Latent aspect rating analysis on review text data: a rating regression approach. In: KDD (2010)
29. Webb, G.I., Pazzani, M.J., Billsus, D.: Machine learning for user modeling. User Model. User Adap. Interact. **11**, 19–20 (2001)

30. Wyllys, R.E.: Empirical and theoretical bases of Zipf's law. Libr. Trends **30**(1), 53–64 (1981)
31. Zhang, W., Wang, J.: Integrating topic and latent factors for scalable personalized review-based rating prediction. IEEE Trans. Knowl. Data Eng. **28**(11), 3013–3027 (2016). https://doi.org/10.1109/TKDE.2016.2598740
32. Zipf, G.K.: Human Behaviour and the Principle of Least Effort. Addison-Wesley, Reading (1949)

Blood Vessel Delineation in Endoscopic Images with Deep Learning Based Scene Classification

Mayank Golhar[1], Yuji Iwahori[2](✉), M. K. Bhuyan[1], Kenji Funahashi[3], and Kunio Kasugai[4]

[1] Department of Electronics and Electrical Engineering,
Indian Institute of Technology Guwahati, Guwahati 781039, India
`mayankgolhar@gmail.com, mkb@iitg.ernet.in`
[2] Faculty of Engineering, Chubu University, Matsumoto-cho 1200,
Kasugai, Aichi 487-8501, Japan
`iwahori@cs.chubu.ac.jp`
[3] Nagoya Institute of Technology, Nagoya 465-8555, Japan
`kenji@nitech.ac.jp`
[4] Department of Gastroenterology, Aichi Medical University, Nagakute-cho,
Aichi-gun, Aichi 480-1195, Japan
`kuku3487@aichi-med-u.ac.jp`
`http://www.cvl.cs.chubu.ac.jp/`

Abstract. A novel blood vessel extraction methodology is proposed in this work. The blood delineation pipeline consists of three stages. In the first stage, a high-level classification of the input endoscopic images is done into four classes, based on the blood vessel information and dye content. For obtaining the classification features two methodologies are used, a ResNet inspired Convolutional Neural Network and a collection of hand picked feature extractors which capture various colour, edge and texture based class information. The features obtained from both are then combined and are fed into an SVM for classification. In the second stage, the classified image containing blood vessel information is then processed with Frangi Vesselness filter for blood vessel extraction. However, it is observed that many non-blood vessel edges are also erroneously detected as blood vessels. To decrease this misdetection, two additions are proposed. One is the dark background subtraction and another is dissimilarity index, which is used to differentiate the non-blood vessel edges from the blood vessel ones. The dissimilarity index, which is another novelty of the paper, exploits the difference of symmetric nature of the blood vessels versus the non-symmetric nature of non-blood vessel edges. The results of the proposed blood vessel delineation algorithm were found to give better accuracy than vanilla Frangi Vesselness filter and BCOSFIRE filter, which is another state-of-the art vessel extraction approach, by 8% and 5% respectively.

Keywords: Endoscopy · Scene classification · SVM · Deep learning
CNN · Blood vessel · Frangi Vesselness

© Springer International Publishing AG, part of Springer Nature 2018
M. De Marsico et al. (Eds.): ICPRAM 2017, LNCS 10857, pp. 147–168, 2018.
https://doi.org/10.1007/978-3-319-93647-5_9

1 Introduction

In areas like retinopathy and endoscopy, the detection and analysis of blood vessels plays a major role in medical diagnosis. For example in diabetic retinopathy, which is a leading cause of blindness among adults, symptoms such as abnormal growth, swelling and leaking of fluid from blood vessels are observed. Early detection of changes in the physiological structure of blood vessels can help in early diagnosis. Similarly in gastroenterology, increase in the size of blood vessels is a symptom of inflammatory bowel diseases like Crohn's disease and Ulcerative Colitis. Whilst, a decrease in the size of vessel, which leads to a reduced blood supply, is a cause of Ischemic bowel diseases like Ischemic Enterocolitis [1].

For segmentation of blood vessels in fundus images an umpteen number of algorithms are available [2]. But, this is not true in the case for endoscopic images. Applying generic segmentation algorithms for intra abdominal images produces egregious result due to the special imaging environment encountered due to high specular reflection, false blood vessel like patterns, more camera sensor noise, deformable colon walls etc. Thus, in this paper we propose a novel algorithm for blood vessel delineation considering the challenging conditions encountered in endoscopic images.

The proposed blood vessel segmentation pipeline consists of two stages, where in the first stage, a scene classification is done based on the blood vessel and ink content using SVM. The feature extraction is done by deep learning and hand crafted feature extractors. In [12], scene classification is done using on hand designed features only. Also, as deep learning is used for feature extraction, the dataset used in this paper is much larger compared to [12]. Earlier, classification of endoscopic images based on tumor texture pattern has been done in [3,4]. But these methods are only applicable on focused images of tumor and they take into account local features. Whereas, the proposed classification uses high level global features for discrimination. The proposed classification is discussed in detail in Sect. 2.1.

The first stage essentially checks at a global level whether blood vessel are present in the input image. In the next stage, the blood vessel containing images are then given to the blood vessel extraction module which segments the blood vessels. Blood vessel segmentation, with an aim for detecting distinctive feature points in colon image has been done in [5]. Here, colon wall's blood vessels' branching points and branching segments are considered as features. Their blood vessel delineation approach is based on Frangi vesselness [6]. This is followed by Ridgeness-based Circle Test and Ridgeness-based Segment Test for detecting branching points and branching segments respectively. But it was observed that the Frangi vesselness method erroneously misclassifies many non blood vessel edges from structures like polyp, colon wall etc. as blood vessel which results in many incorrect feature points. Thus, in [12] a selection of techniques such as dark background subtraction and dissimilarity index, are suggested to improve the performance of blood vessel segmentation by removing edges obtained from non-blood vessel elements. The procedure proposed is invariant to illumination, scaling and orientation. Details about these are given in Sects. 2.3 and 2.4.

2 Proposed Method

The approach is briefly summarized in the Fig. 1. The blood vessel extraction pipeline consists of two major modules. In the first module, a high level classification of image is done to ascertain whether the scene contains blood vessel information or not. In the second module, the classified blood vessel containing image is initially pre-processed to remove noise and specular components. Then, the Frangi Vesselness algorithm is applied to segment the blood vessels. The pre-processed image is also used to do background subtraction. After computing the vesselness image, the dissimilarity index of edges is calculated by the dissimilarity detecting filtering procedure, to determine whether they are obtained from a blood vessel or not. The non blood vessel edges are dropped in the final result. The above steps are explained in detail in the following sections and in Fig. 2.

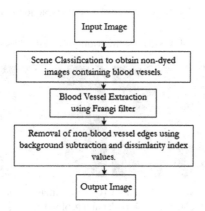

Fig. 1. Overview of proposed blood vessel extraction algorithm.

2.1 Scene-Based Classification

Motivation. During a live colonscopy examination, a dynamically changing environment is encountered. The endoscopic video of the examination will contain images of colon where physiological structures like polyps, blood vessels etc. may or may not be present in all frames. Not only this, during the examination various diagnostic activities take place such as surgical removal of polyp, dyeing or tattooing the colon with ink for better visualisation of polyps as well marking smaller polyps for future reference, washing of colon with medicinal liquid etc. In this potpourri of scenes, it is essential to select only those frames in which blood vessels are present.

This assortment of scenes is classified into four categories for our purpose of blood vessel extraction as follows:

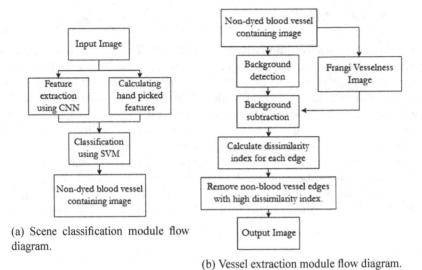

(a) Scene classification module flow diagram.

(b) Vessel extraction module flow diagram.

Fig. 2. Detailed flow charts of scene classification and blood vessel delineation modules.

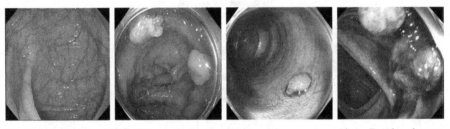

(a) Non-dyed images containing blood vessel - Class 1

(b) Non-dyed images not containing blood vessels - Class 2

(c) Dyed images containing blood vessels - Class 3

(d) Dyed images not containing blood vessels - Class 4

Fig. 3. Representative images of Classes 1 to 4.

Class 1: Non-dyed images containing blood vessel.
Class 2: Non-dyed images not containing blood vessels.
Class 3: Dyed images containing blood vessels.
Class 4: Dyed images not containing blood vessels.

The classification into dyed and non dyed images is necessary as both these types would require different vessel segmentation techniques. This is because in ink images, special conditions are encountered, like higher specular reflections, ink pattern texturally similar to blood vessel pattern, etc. [12] discusses vessel segmentation in non-dyed images only. Delineation of blood vessel in dyed images is a topic for future work. Illustrative images for the classes are shown in Fig. 3.

For feature extraction from endoscopic images, two modalities were employed. Firstly, features were hand-picked to capture the color, edges and texture content information. The features were then used to train Support Vector Machine (SVM) model, as done in [12]. In this paper, additional approach of deep learning is used. Convolutional Neural Network (CNN) was used for the purpose of classification as well as for feature extraction. The features extracted by CNN were then used to train another SVM model. Finally the featured extracted by CNN and the hand picked features were concatenated and again used to train another SVM model. A detailed discussion is done in the following section.

Methodology. Defining a feature vector to discriminate between the required classes is a challenging task. For feature selection, [7,8] were referred. In [12], the feature vector was defined on the basis of the following three criteria:

1. **Colour-based Features:** The non dyed images have more red component whereas the dyed images have more blue colour component. So, colour based statistical features can be used for classification. For every color, the 10 - bin first order histogram values, the mean, variance, skew, energy and entropy are used as features. Thus, the total number of colour-based features used were 45.

2. **Edge-based Features:** In general, it is observed that the blood vessel containing classes have more edges compared with blood vessel absent classes. Thus, edge based statistical features can be used for classification. The Canny edge operator is used on every channel for getting the gradient magnitude and direction. For every color, 10 - bin histogram values, the mean, variance, skew, energy and entropy derived from gradient magnitude are used as features. A histogram of directional angles with central bin values $\{-90°, -45°, 0°, 45°, 90°\}$ is constructed. The histogram bin counts, the mean, variance, skew, energy and entropy obtained from directional angle values are also used as features. The total number of edge-based features used were 75.

3. **Texture-based Features:** The texture of blood vessel edges is different from texture from edges due to colon walls, polyps etc. Thus, the following texture information capturing statistical features are proposed. Image is first converted to grayscale and then:

   (a) The Fast Fourier Transform (FFT) transform of the image is calculated. The mean, variance, skew, energy, entropy of the FFT and grayscale images are used as features. Also the range of each column of grayscale image is used as feature.

   (b) A gabor filter bank is created, with filters having different wavelengths and orientations. The orientations of gabor filter used are $\{0°, 45°, 90°, 135°\}$ and wavelengths increasing exponential, in range $[2\sqrt{2}, \text{Hypotenuse of image}]$. For each filtered image, the 10 - bin histogram values, the mean, variance, skew, energy and entropy are used as features. The total number of texture-based features were 210.

For the sake of completeness, the formulae used above are given. The first order histogram probability is given by:

$$P(g) = N(g)/n; \qquad (1)$$

where $N(g)$ the bin count of the gth histogram level and n is the total number of pixels.

$$Mean(\mu) = \frac{\sum_{i=1}^{n} x_i}{n} \qquad (2)$$

where x_i is intensity of the pixel of a particular channel.

$$Standard\,deviation(\sigma) = \sqrt{\frac{\sum_{i=1}^{n}(x_i - \mu)^2}{n}} \qquad (3)$$

$$Skew = \frac{1}{n}\frac{\sum_{i=1}^{n}(x_i - \mu)^3}{\sigma^3} \qquad (4)$$

$$Energy = \sum_{i=1}^{n} x_i^2 \qquad (5)$$

$$Entropy = -\sum_{g=1}^{L} P(g)log(P(g)) \qquad (6)$$

where L is the total number of histogram levels.

The above features are then used to train a classifier which in our case is a SVM. Support Vector Machine is used as classifier as the aim was to make classification robust but computationally less intensive. As SVM is a binary classifier and we have total four classes, One V/s One classification was used. As the data is not necessarily linearly separable, different kernels were used. It was observed that cubic kernel gave the best results. Results are explained in detail in Sect. 3.1.

Convolutional Neural Network. Current trends have shown the rise of CNNs as an emerging choice among researchers as image classifiers and feature extractors. It is observed that in comparison to hand picked feature extraction algorithms, CNNs [9] provide more robustness in terms of extracting highly discriminative global and local features. In CNNs, there are multiple convolutional layers which are followed by pooling layers which help in dimensionality reduction as well as inducing geometric invariance to translation, rotation etc. The kernels in the convolutional layers are trained as feature extractors. In our experiments, we have used CNN both as a classifier and feature extractor. The details of experiments are described in Sect. 3.1. The details of CNN architecture used and the salient points of network are described in the following section.

CNN Architecture. Among the many CNN models which were experimented with, the following ResNet [10] inspired model was found to give the best results as a classifier. The architecture is described as follows.

The network consists of four residual learning blocks followed by three fully connected layers. A residual learning block consists of a convolutional layer with 64 filters of size $3 \times 3 \times 3$ with stride of 1, followed by ReLU activation function. The output of Relu is fed into another convolutional layer with 64 filters of size $3 \times 3 \times 3$ and stride 1. Its output is then added to a feed forward shortcut connection which carries an identity mapping from the input of the residual learning block. This is then max pooled by a non-overlapping kernel of size 2×2 and stride 2. Figure 4 shows a diagrammatic representation of a residual learning block.

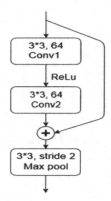

Fig. 4. Residual learning block.

After four such residual blocks, the output of the final residual block is then connected to the fully connected layer with 1024 number of neutrons which are further connected to 2nd fully connected layer with 64 neurons and finally to the output with four nodes, each one represents the probability of being in that particular class. The complete architecture of the net used is shown in Fig. 5.

Each convolved as well as fully connected layer output is passed through an activation function. In our model, RELU was used as activation function. The output of last layer is passed through softmax layer to convert the output into probabilities. The labels of each classes are encoded into one-hot vectors which are helpful while comparing the probability output as well as calculating the cross entropy loss. In the loss function, L_2 regularisation loss was also added for each filter to avoid overfitting. Dropout [11], another regularisation technique where neurons are dropped with some probability to give an effect of model averaging, was incorporated. A keep probability of 0.6 was used. For optimization, gradient descent gives decent results but many times it gets stuck at local minima. To overcome this, we need to consider the momentum at which the weights are getting modified. Also, an adaptive learning rate is required for optimal convergence to global minima. All these issues are addressed by using Adam-optimizer. Weights are initialized as normal random variable with zero-mean and standard deviation equal to $\sqrt{2/Fan_{in}}$ where fan_{in} is the number of neurons in the filter. The batch size in each iteration is taken as 25 images. Due to limited dataset, augmentation on images is performed by taking overlapping crops of images.

To use CNN as feature extractor, the fully connected layers were removed and the output of the final residual block was used as feature for the corresponding input image. Using CNN a 127×127 image was reduced to a 192-dimension vector. An SVM model was trained using these features. To further boost the discriminative ability, these CNN features were combined with the hand engineered features and then used to train an SVM. Detailed comparison of these four models is given in Sect. 3.1.

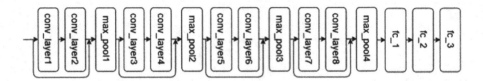

Fig. 5. Complete CNN architecture.

2.2   Blood Vessel Extraction

The endoscopic images are known to have high specular reflection component due to the mucosal lining of the colon. To remove the specular reflection, the algorithm suggested in [13] was used. The detected specular reflections are then treated as no information regions. The proposed blood vessel algorithm in [12] is based on the papers [5,6]. The Frangi vesselness filter is used to segment the blood vessels. The Vesselness image is then skeletonised, as mentioned in [5] to obtain the Ridgeness image. In the Frangi vesselness image, edges from non-blood vessel structures like polyps are incorrectly identified as blood vessel edges.

For the completeness of paper, their approach is briefly summarized as follows:

1. Pre-Processing
   (a) The green channel of the preprocessed image is used, as it gives the best contrast between the background and the vessels.
   (b) Convert to scale space model.
   (c) Calculate Hessian matrix for each point.
   (d) From the Hessian matrix, eigenvalues and eigenvectors are calculated, say λ_1 and λ_2. The points having λ_1 - Medium value and λ_2 - High value are considered as a candidates of blood vessels.
2. Blood vessel enhancement
   (a) Parameter ridgeness and vesselness are defined as

$$Vesselness(\sigma) = exp(\frac{\lambda_1^2/\lambda_2^2}{2\beta^2}).(1 - exp(\frac{-(\lambda_1^2 + \lambda_2^2)}{2c^2})) \qquad (7)$$

$$Ridgeness(x, y, \sigma) = Vesselness(x, y, \sigma).$$

$$abs\{sign(\nabla I(x + \epsilon u_2, y + \varepsilon v_2, \sigma)) - sign(\nabla I(x - \epsilon u_2, y - \varepsilon v_2, \sigma))\}/2, \tag{8}$$

where in the ridgeness formula u_2 and v_2 are the x and y components of the eigenvector pointing in the direction perpendicular to that of blood vessel. In the vesselness formula, β and c are soft thresholds. Ridgeness for all the candidate points are calculated. Thereafter, the pixels which have local maximum are only retained.

(b) Single pixel width ridges, representing the blood vessel skeleton are obtained as a result. Background noise is also removed.

The further part of the section explains in detail the proposed methods, namely background subtraction and calculation of dissimilarity index, to reduce this error.

2.3 Background Removal

Source of Error. When the orientation of endoscopic camera is perpendicular to colon wall, the result of blood vessel extraction of such images is found to be acceptable. Whereas, if the camera is oriented parallel to the wall of colon, far away regions devoid of illumination appear as dark background in the image. In this dark background, many false noisy edges are detected as blood vessels. An example given in [12] is shown in Fig. 6. So, dark background is segmented to remove this error.

Methodology. Various methods were used to segment the dark background namely, Otsu's single level thresholding, k-means clustering and Otsu's multi level thresholding, for each of the three colours. Among these it was experimentally found that *Otsu's multi-level thresholding with two thresholds using Red channel* gave the best results. A point to be noted is that instead of using a single threshold which would have resulted in two clusters, we are using two thresholds which results in forming three clusters. The cluster of pixels with the lowest intensity is labeled as background region and the other two are considered as the bright foreground region.

When the camera is parallel to the wall of colon, the image can generally be divided into three different regions based on illumination. Brightly illuminated region is found close to the light source which is mounted on the camera, mid range regions have moderate illumination whereas the poor illumination is found in the regions far away from the source. This observation provided a motivation to partition the image into three clusters rather than two. Because of sufficient illumination, the close by and mid range regions are found to have useful clinical information but faraway regions, due to lack of illumination were found not to contain much interpretable information and therefore were discarded.

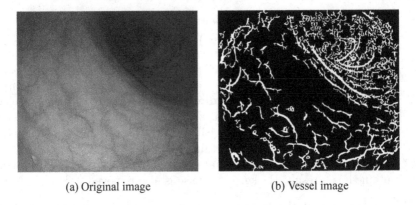

(a) Original image   (b) Vessel image

Fig. 6. Noisy edges in dark background region identified as blood vessels.

2.4   Removing Non-blood Vessel Edges

Source of Error. In the result of Frangi vesselness, it is found that the edges from polyps, ridges or folds of wall of colon, medical suture and specular reflections are also detected as blood vessels. This section discusses in detail a method proposed in [12], to distinguish between the edges from blood vessels and other non blood vessel edges. A representative image given in [12] showing the above error is given in Fig. 7.

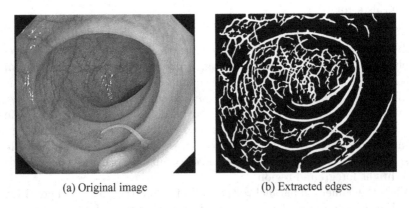

(a) Original image   (b) Extracted edges

Fig. 7. The ridges of inner walls, polyp and medical suture have been erroneously extracted as blood vessels in vesselness image.

Motivation. Careful observation reveals that the edges obtained from non blood vessel structures like polyps, medical suture, ridges etc. are fundamentally different from the edges obtained from blood vessel. This is evident if the colour intensities of neighbourhood regions of the edges are considered. The differences are as follows:

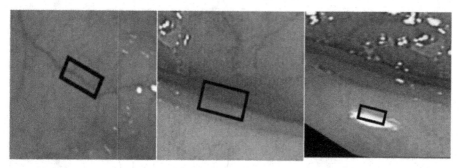

(a) Window around blood vessel   (b) Window around ridge   (c) Window around specular component

Fig. 8. Windows around various structures, showing the differences in intensity in the two halves of window.

1. **Blood Vessel:** If a rectangular window is placed such that the center line of the blood vessel divides the window into two halves, it is found that intensities of the two halves are very similar as illustrated in Fig. 8a in [12].
2. **Ridges, Polyp Edges, Medical Suture and Specular Components:** Similarly if a window is placed on the edge of a ridge, specular component, medical suture or polyp, with the edge dividing the window into two halves, it is found that intensities of the two halves are very dissimilar. An example given in [12] is shown in Fig. 8b and c, one side is brighter whereas the other side is darker. It is also observed that the vesselness value of these structures is higher than the blood vessels.

Overview of the Approach. Based on the lines of motivation above, we can distinguish between blood vessel and non blood vessel edges by exploiting the property of intensity difference in the two halves. So, a metric called dissymmetry index is defined which quantifies the intensity difference between the two halves. A higher dissymmetry index indicates wider intensity difference in the halves indicating the edge is not a blood vessel. The dissymmetry index for each edge is calculated by a dissymmetry detecting filtering procedure. The formula of dissymmetry index and filtering procedure as given in [12] are as follows:

1. As it is found that the vesselness value of non blood vessel edges is higher than the blood vessel edges, an initial thresholding is done. After thresholding, mostly the edges from polyps, medical suture, ridges, specular components remain. Though most of the blood vessel edges are eliminated, some remain.
2. The vesselness image is then converted to the *ridgeness* image, which is the skeletonised version of vesselness image. This is done to obtain the center lines of the edges.
3. After finding the ridgeness image, the custom intensity-based dissymmetry detecting filtering is done. This filter emulates the role of the window

described in the previous section. The filter is a rotating filter which is always oriented in the direction of blood vessel. The minor eigenvector calculated from the Hessian matrix at the center of filter gives the direction of the blood vessel. Like the windows previously described, the rotating filter is positioned such that the center line of the edge coincides with that of filter.

$$f(i,j) = \underset{\theta}{rotate}\frac{1}{25}\begin{pmatrix} -1 & -1 & -1 & -1 & -1 \\ -1 & -1 & -1 & -1 & -1 \\ 0 & 0 & 0 & 0 & 0 \\ 1 & 1 & 1 & 1 & 1 \\ 1 & 1 & 1 & 1 & 1 \end{pmatrix} \qquad (9)$$

where θ represents the direction of the minor eigenvector, which is along the direction of edge. Another point to be noted is that, due to lack of space the filter size shown is 5×5 pixels but in actual implementation window size of 20×20 pixels was used.

4. The filter is placed on the pixels lying on the center line of the edge in the ridgeness image. For filtering, the Gaussian smoothened red channel of the original image is used as input. The function of the filter is to find the sum of intensities of pixels lying in each half and return the absolute difference of these sum-of-intensities of the two halves.

$$g(i,j) = \begin{cases} abs|f(i,j) * x(i,j)| & \text{if } x(i,j) \in \text{ Center} \\ 0 & \text{otherwise.} \end{cases} \qquad (10)$$

where $f(i,j)$ is the custom filter, $x(i,j)$ is the gaussian filtered original image's red channel and $g(i,j)$ is the filtered image. It is to be noted that $g(i,j)$ is calculated for center line pixels of edges only.

5. For an edge, the Dissymmetry Index is average of all the filtered values of the center line pixels lying inside that edge. Mathematically, Dissymmetry index of the kth edge is given by

$$Dissymmetry\, index(k) = \frac{1}{|S_k|} \sum_{g(i,j) \in S_k} g(i,j) \qquad (11)$$

where S_k is the set of all pixels of filtered image belonging to the kth edge. $|S_k|$ is the total number of pixels belonging to the kth edge.

6. Now thresholding is done based on the dissymmetry index value. All the edges with dissymmetry index above the threshold, indicating they have contrasting halves, are marked as non-blood vessel edges.

7. The steps 4 to 7 are repeated using the green channel of the original image as input. Filtering using both red and green channel is done as it is found that some non-blood vessel edges are better captured in the red channel whereas others in the green channel.

8. The results obtained from the red and green channel are then OR'ed together. The resulting image consists mostly of the unwanted specular components, polyps, medical suture and other non-blood vessel edges.

9. This OR'ed image is then subtracted from the vesselness image to remove the error. The final resulting image will consists mostly of blood vessels.

3   Experiments and Results

3.1   Scene Classification

For the purpose of scene classification, four different models were used, SVM trained on hand crafted features, CNN used as image classifier, SVM trained on CNN extracted features and SVM trained on combination of hand engineered and CNN extracted features. In this paper, the data set used for training and testing in all models comprised of a total of 38,204 images with 8741, 9146, 10997 and 9320 in Classes 1, 2, 3 and 4 respectively. Initially, the class 4 contained only 4660 images, which would have caused the problem of data imbalance among classes. To navigate this, data augmentation was done, where in every image, two overlapping crops each occupying 80% of the original image were taken. All the images were then resized to $127 \times 127 \times 3$. For giving class labels, data set images were manually classified by visual inspection into their respective classes. The dataset of scene classification which earlier consisted on 513 images in [12], was increased to 38,204 images in this paper. Earlier an SVM based model was tried out where the features were hand picked (pre-designed feature extractors). This model gave an accuracy of 85.4%. When the newer dataset was used for training this model, the accuracy jumped to 98.1%. Now 3 more newer models were experimented with. Also, as a result of using deep learning features, the accuracy of the final model was much higher in this paper.

Each model is described in detail as follows:

1. **SVM Trained on Hand Crafted Features:** As there is no guarantee that the data is linearly separable, different kernels were experimented with, to find out which gave the best results. Linear, quadratic, cubic, medium Gaussian and coarse Gaussian kernels were used. The results obtained for the kernels are shown in Table 1. The One versus One classification was used as it gave comparatively better results than One versus All classification. For measuring the predictive performance of the statistical model, holdout validation was done with 20% holdout. The *One Vs One SVM with cubic kernel gave the best result with an accuracy of 98.1%*. It was observed that if dimensionality reduction is done using Principal component analysis (PCA), the accuracy fell by around 1%, if the variance retained is 99%. The post PCA results are shown in the Table 1. The confusion matrix depicting the true positive rate and false negative rate is shown in Table 2.

2. **CNN as Image Classifier:** The CNN architecture described in Sect. 2.1 was used for classification of images. A training accuracy of 97.9% was obtained. Holdout validation with 20% was used for testing. *The validation accuracy obtained was 97.2%*. Generally, the performance of CNN as a classifier is better than linear classifier like SVM trained using hand engineered features. But in this case, the CNN accuracy is found to be lower. One of the reason is that CNNs generally require large dataset for training properly as compared to SVM. This is one of the challenges faced while working with medical data, as publicly annotated medical datasets are not easily available.

3. **SVM Trained on CNN Features:** In this model, the fully connected layers are removed and output of the last residual learning block is taken as the extracted feature of the input image, thereby using CNN as a feature extractor. These features are then used to train SVM model. As the features extracted by the convolutional layers are still not linearly separable, different kernels are used for transformation. In this model also, we used holdout validation with 20% holdout. The *One V/s One SVM with Cubic Kernel gave the best result with 97.6% validation accuracy* as shown in Table 3. Post PCA, the accuracy fell by 0.5–1% when 99% energy was retained. The confusion matrix for the cubic kernel is given in Table 4.

4. **SVM Trained on Combination of CNN and Hand Picked Features:** To further improve upon the result, feature vectors obtained from CNN and the hand crafted features were concatenated to form a 518 dimensional vector. This was then passed to an SVM model for training with different kernels. Again it is observed that the cubic kernel gave the best result. Among all the four model, *the One V/s One SVM with cubic kernel trained on CNN extracted and hand picked features gave the best result with an accuracy of 99.2%* as shown in Table 5. This is on expected lines, as combining features taps the strengths of both the methods. It is highly possible that the class information which might have been missed out while designing hand crafted features, could have been expressed in the CNN features. Even post PCA, the accuracy only fell by 0.1% when 99% of the variance was retained. This tells that the features used, were highly discriminative and uncorrelated. The confusion matrix is given in Table 6.

Table 1. Accuracy of SVM trained on hand crafted features.

Kernel	Accuracy (without PCA)	Accuracy (post PCA)
Linear	94.3%	93.6%
Quadratic	97.5%	97.0%
Cubic	98.1%	97.2%
Medium Gaussian	97.4%	96.0%
Coarse Gaussian	93.0%	91%

3.2   Background Removal

To do background subtraction, various techniques were used in [12]. The results obtained for each methodology are discussed below:

1. **Adaptive Thresholding using Otsu's Method: Single Threshold:** The popular Otsu's method was used for determining the global threshold. Otsu's method tries to find the threshold by minimizing intra-class variance. However, for most of the images the results obtained were not accurate. An example image given in [12] where it failed is shown in Fig. 9.

Table 2. Confusion matrix for SVM trained on Handcrafted features with cubic kernel.

True class	Predicted class			
	Class 1	Class 2	Class 3	Class 4
Class 1	94%	>5%	0%	<1%
Class 2	<1%	99%	<1%	<1%
Class 3	<1%	<1%	99%	<1%
Class 4	0%	1%	1%	98%

Table 3. Accuracy of SVM trained on CNN extracted features.

Kernel	Accuracy (without PCA)	Accuracy (post PCA)
Linear	96.1%	95.8%
Quadratic	97.4%	97.2%
Cubic	97.6%	97.2%
Medium Gaussian	97.5%	96.9%
Coarse Gaussian	95.6%	94.8%

Table 4. Confusion matrix for SVM trained on CNN extracted features with cubic kernel.

True class	Predicted class			
	Class 1	Class 2	Class 3	Class 4
Class 1	94%	5%	<1%	<1%
Class 2	2%	97%	<1%	<1%
Class 3	<1%	<1%	99%	<1%
Class 4	0%	1%	1%	98%

Table 5. Accuracy of SVM trained on combination of CNN extracted and hand crafted features.

Kernel	Accuracy (without PCA)	Accuracy (post PCA)
Linear	98.9%	98.7%
Quadratic	99.1%	99.0%
Cubic	99.2%	99.1%
Medium Gaussian	99.1%	98.8%
Coarse Gaussian	98.3%	97.7%

Table 6. Confusion matrix for SVM trained on combination of CNN extracted and hand picked features with cubic kernel.

True class	Predicted class			
	Class 1	Class 2	Class 3	Class 4
Class 1	98%	2%	0%	0%
Class 2	<1%	99%	<1%	0%
Class 3	0%	<1%	>99%	<1%
Class 4	0%	0%	1%	99%

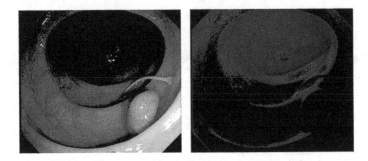

Fig. 9. Adaptive single threshold based clustering did not give acceptable result.

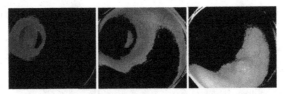

(a) Result of clustering using Red channel values.

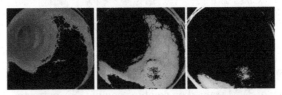

(b) Result of clustering using Green channel values.

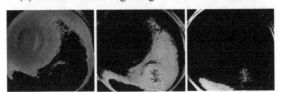

(c) Result of clustering using Blue channel values.

Fig. 10. Results of K-means clustering using RGB.

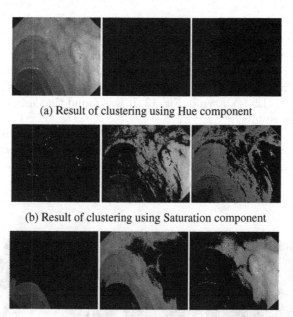

(a) Result of clustering using Hue component

(b) Result of clustering using Saturation component

(c) Result of clustering using Value component

Fig. 11. Results of K-means clustering using HSV.

2. **K-means Clustering:** Another popular technique for segmentation of dark background and bright foreground is the K-means clustering. Clustering was done on different channels of different colour spaces like RGB, CIE-Lab etc. to find out which gave the best results. A comparison is given among all the tested color spaces and channels.

   K-means clustering was done into three clusters using initial seeds as 0.1, 0.5 and 0.75. Though it was observed that the choice of initial seeds did not affect the performance. K-means++ also can be used to determine the initial choices of centers.

   (a) *Comparison between R, G and B Color Channels:* An illustrative comparison of clustering with Red, Blue and Green colors separately is shown in Fig. 10 as given in [12]. It was found that Red color channel gave the best results for segmentation in most cases.

   (b) *Comparison between H, S and V Channels:* In [12], an illustrative comparison of clustering with H, S and V channels is shown in Fig. 11. It was observed that V channel of the image gave the best results. As expected, it's clustering result was similar to the R channel.

   (c) *Comparison between L* and a*b* Color Channels:* The L*a*b* color space is derived from the CIE XYZ tristimulus values. The L*a*b* space consists of a luminosity 'L*' or brightness layer, chromaticity layer 'a*' indicating where color falls along the red-green axis, and chromaticity layer 'b*' indicating where the color falls along the blue-yellow axis.

Comparison between L* and a*b* space, given in [12] is shown in Fig. 12. The best segmentation was done by using L* channel.

(d) *Comparison between R and L* Color Channels:* Both the RGB and CIE Lab color space gave good results using the 'R' and 'L*' channels, respectively. Thus, a comparison is made to between the 'R' and 'L*' channel as given in [12] is shown in Fig. 13. It was found that using the R channel gave best results.

(e) *Adaptive thresholding using Otsu's method using 2 thresholds:* Otsu's method was used to find two thresholds resulting in three clusters with the Red Channel as the input. The results obtained by k-means clustering and Otsu's multi-level thresholding were found to be almost identical in [12], as shown in Fig. 14.

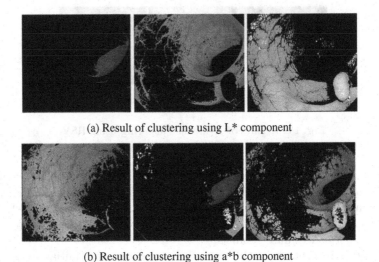

(a) Result of clustering using L* component

(b) Result of clustering using a*b component

Fig. 12. Results of K-means clustering using Lab.

The results of Otsu's 2 level adaptive thresholding and k-means clustering were found to be identical. Hence, considering the fact that k-means clustering being an iterative method is generally slower, it can be concluded that *Otsu's multi-level thresholding using two thresholds on R channel* is the best, among all investigated methods, for dark background segmentation in endoscopic images. Another interesting observation is that when there are no dark regions, the clustering automatically detects two clusters with none labeled as background.

3.3   Removal of Non-blood Vessel Edges

In [12], two techniques are proposed to eliminate the non-blood vessel edges. First was background subtraction and another is using the dissimilarity index, which quantifies whether the edge belongs to blood vessel or not.

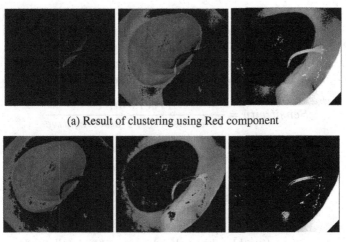

(a) Result of clustering using Red component

(b) Result of clustering using L* component

Fig. 13. K-means clustering result comparison of Red and L* components.

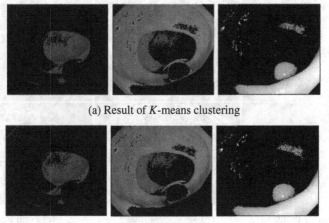

(a) Result of K-means clustering

(b) Result of Adaptive thresholding using Otsu's method with 2 thresholds

Fig. 14. K-means clustering and adaptive thresholding using Otsu's method with 2 thresholds gave similar results.

A dataset of 61 images was used for evaluation. For ground truth, the blood vessel were manually marked in all the images. The values of various thresholds used in evaluation are: Strong edge *vesselness* threshold = 0.9, Green Channel-Dissimilarity Index threshold = 3 and Red Channel-Dissimilarity Index threshold = 3. The values used are found empirically to give the best results on the test dataset. The comparison of various algorithms' results obtained in [12] is shown in Table 7.

The proposed method was found to give better results than both the vanilla Frangi Vesselness method and BCOSFIRE filter. It performed better than BCOSFIRE filter [14], a popular technique in vessel delineation, by around 50% in terms of sensitivity. In terms of the overall accuracy, the proposed method's result is around 5% better than the BCOSFIRE result.

$$Sensitivity = \frac{TruePositive}{TruePositive + FalseNegative} \qquad (12)$$

$$Specificity = \frac{TrueNegative}{Truenegative + FalsePositive} \qquad (13)$$

$$Accuracy = \frac{TruePositive + TrueNegative}{N} \qquad (14)$$

where N is the total number of pixels. Sensitivity is a measure of how good the algorithm was at detecting blood vessels while specificity tells us how correctly could the algorithm identify the non-blood vessel pixels.

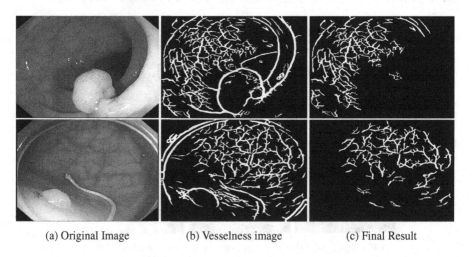

(a) Original Image   (b) Vesselness image   (c) Final Result

Fig. 15. Comparison of input, vesselness and final result.

The proposed method performed better in terms of both the accuracy and specificity by around 8% and 4% respectively, when compared with the vanilla Frangi vesselness. This verifies that the proposed method's focus on removing falsely identified edges as blood vessel edges was correct. The sensitivity of Frangi's method was better by around 3.5%. The reason for this is, while removing edges, some true blood vessel edges were also removed along with false ones. This is tolerable as our end goal is to get correct blood vessel information. This can be obtained from the other detected true blood vessels. Illustrative images with results improved by the proposed method in [12] are shown in Fig. 15.

Table 7. Comparison of results of various vessel delineation methods.

	BCOSFIRE	Frangi's vesselness	Proposed method
Sensitivity	20.22%	75.24%	71.77%
Specificity	93.65%	86.16%	94.57%
Accuracy	88.68%	85.42%	93.03%

4   Conclusion and Future Work

In this paper, a classification of endoscopic images based on scene information was attempted. The classes were organised on the basis of blood vessel information and dye content. Feature selection was done by using hand designed feature extractors and CNNs. The hand engineered feature extractors capture the texture, edge and color information of the image. A ResNet inspired architecture was used for CNN, which acted as feature extractor. Both these features were then used to train an SVM with a cubic kernel. Post classification, only the blood vessel containing images are then processed for detailed blood vessel extraction. The delineation approach is based on the Frangi Vesselness filter. The novelty of the paper lies in the proposal of two techniques to reduce the error of detection of Frangi filter. One is the dark background removal and another is the dissimilarity index which has the ability discriminate between the blood vessel and non-blood vessel edges.

Development of blood vessel segmentation method for dyed images is a topic of future work. A larger dataset can be created and annotated for classification using CNN. A future application of this work is using blood vessels' branching points as feature points for 3D recovery of absolute shape and size of polyps.

Acknowledgment. Iwahori's research is supported by Japan Society for the Promotion of Science (JSPS) Grant-in-Aid Scientific Research(C) (#17K00252) and Chubu University Grant.

References

1. Kumar, V., Abbas, A.K., Fausto, N., Aster, J.C.: Robbins and Cotran Pathologic Basis of Disease. Elsevier Health Sciences, New York (2014)
2. Fraz, M.M., Remagnino, P., Hoppe, A., Uyyanonvara, B., Rudnicka, A.R., Owen, C.G., Barman, S.A.: Blood vessel segmentation methodologies in retinal imagesa survey. Comput. Methods Programs Biomed. **108**(1), 407–433 (2012)
3. Tamaki, T., et al.: Computer-aided colorectal tumor classification in NBI endoscopy using local features. Med. Image Anal. **17**(1), 78–100 (2013)
4. Hafner, M., Gangl, A., Wrba, F., Thonhauser, K., Schmidt, H.-P., Kastinger, C., Uhl, A., Vecsei, A.: Comparison of k-NN, SVM, and NN in pit pattern classification of zoom-endoscopic colon images using co-occurrence histograms. In: ISPA 2007 5th International Symposium on Image and Signal Processing and Analysis. IEEE (2007)

5. Lin, B., Sun, Y., Sanchez, J.E., Qian, X.: Efficient vessel feature detection for endoscopic image analysis. IEEE Trans. Biomed. Eng. **62**(4), 1141–1150 (2015)
6. Frangi, A.F., Niessen, W.J., Vincken, K.L., Viergever, M.A.: Multiscale vessel enhancement filtering. In: Wells, W.M., Colchester, A., Delp, S. (eds.) MICCAI 1998. LNCS, vol. 1496, pp. 130–137. Springer, Heidelberg (1998). https://doi.org/10.1007/BFb0056195
7. Mohanty, A.A., Vaibhav, B., Sethi, A.: A frame-based decision pooling method for video classification. In: 2013 Annual IEEE India Conference (INDICON). IEEE (2013)
8. Zhang, D., Wong, A., Indrawan, M., Lu, G.: Content-based image retrieval using Gabor texture features. In: IEEE Pacific-Rim Conference on Multimedia, University of Sydney, Australia (2000)
9. Alex, K., Sutskever, I., Hinton, G.E.: Imagenet classification with deep convolutional neural networks. In: Proceedings of the Advances in Neural Information Processing Systems 25 (NIPS 2012), pp. 1097–1105. Curran Associates, Denver (2012)
10. He, K., Zhang, X., Ren, S., Sun, J.: Deep residual learning for image recognition. In: Proceedings of the IEEE Conference on Computer Vision and Pattern Recognition, pp. 770–778 (2016)
11. Srivastava, N., Hinton, G.E., Krizhevsky, A., Sutskever, I., Salakhutdinov, R.: Dropout: a simple way to prevent neural networks from overfitting. J. Mach. Learn. Res. **15**(1), 1929–1958 (2014)
12. Golhar, M., Iwahori, Y., Bhuyan, M.K., Funahashi, K., Kasugai, K.: A robust method for blood vessel extraction in endoscopic images with SVM-based scene classification. In: Proceedings of the 6th International Conference on Pattern Recognition Applications and Methods, vol. 1, pp. 148–156. ICPRAM, (2017). https://doi.org/10.5220/0006192601480156. ISBN 978-989-758-222-6
13. Ikeda, N., Usami, H., Iwahori, Y., Kijsirikul, B., Kasugai, K.: Generating Lambertian image by removing specular reflection component and difference of reflectance factor using HSV. In: Proceedings of the ITC-CSCC 2016, T2-5, Computer Vision (2), pp. 547–550 (2016)
14. Azzopardi, G., Strisciuglio, N., Vento, M., Petkov, N.: Trainable COSFIRE filters for vessel delineation with application to retinal images. Med. Image Anal. **19**(1), 46–57 (2015)

Semi-automated Testing of an Architectural Floor Plan Retrieval Framework: Quantitative and Qualitative Comparison of Semantic Pattern-Based Matching Approaches

Qamer Uddin Sabri[1,2], Johannes Bayer[1,2]([✉]), Viktor Ayzenshtadt[1,3],
Syed Saqib Bukhari[1,2], Klaus-Dieter Althoff[1,3], and Andreas Dengel[1,2]

[1] German Research Center for Artificial Intelligence,
Trippstadter Strasse 122, 67663 Kaiserslautern, Germany
{Qamer_Uddin.Sabri,Johannes.Bayer,Viktor.Ayzenshtadt,Saqib.Bukhari,
Klaus-Dieter.Althoff,Andreas.Dengel}@dfki.de
[2] Technical University Kaiserslautern, P.O. Box 3049,
67663 Kaiserslautern, Germany
[3] Institute of Computer Science, University of Hildesheim,
Samelsonplatz 1, 31141 Hildesheim, Germany
http://www.dfki.de

Abstract. Early design phases in architecture deal with the conceptualization of a building. During these phases, a high-level description of a building (usually coming from a contractor of costumer) is iteratively turned into a first floor plan layout. One established method for architects to get inspiration is the search of references from former building projects. However, this search is usually conducted manually (and therefore labor-intensive) nowadays. Hence, an automated search for similar architectural concepts is desired. In the course of this paper, case-based reasoning and (in)exact graph matching are utilized to construct an end-to-end system for floor plan retrieval, accessible by a refined version of our design-supporting web interface. In our approach, a floor plan is modeled as a graph, where each room is represented as a node and the relations between rooms are modeled as edges. We use a set of high-level abstractions, so-called semantic fingerprints, to generate simplified graphs that are simple to match. The retrieval process itself is performed by three systems (case-based reasoning, exact graph matching and inexact graph matching), whose results are unified internally. We conducted several tests to show the deployment ability of our system: firstly, we run a stress-test for determining the computational limits our system can handle. Secondly, we tested our system qualitatively and showed that each retrieval system is superior in at least one search scenario.

This paper is an extended version of [1]. In the paper at hand, we introduce a new feature that maps components of search queries to results

---

Q. U. Sabri, J. Bayer and V. Ayzenshtadt have equally contributed to the paper.

M. De Marsico et al. (Eds.): ICPRAM 2017, LNCS 10857, pp. 169–189, 2018.
https://doi.org/10.1007/978-3-319-93647-5_10

and demonstrate this function by the means of a case study. Finally, we conducted an extended literature comparison of the case-based system in this area.

Keywords: Graph matching · Subgraph matching
Graph isomorphism · Architectural floor plan · Case-based reasoning
Pattern recognition · Query-result mapping

1   Introduction

During the early phases of architectural design, the architect's task is to develop a first, rough floor plan layout given a high-level description of the building. In order to accomplish this task, different working methods have been established. In general, working with references of previously completed building projects is common. However, searching such references is usually conducted manually nowadays, involving the labor-intensive and manual consultation of dedicated magazines and libraries. Speeding up this process by computerized means is therefore desired. To address this issue, we have already introduced MetisCBR [4], an approach for distributed case-based retrieval of similarly structured floor plans.

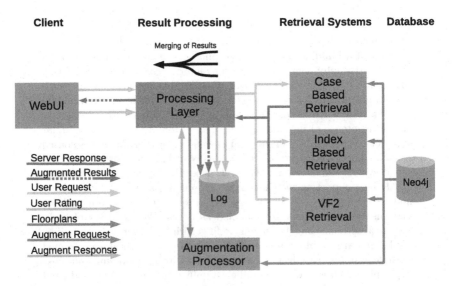

Fig. 1. Overview over the system architecture of Archistant (simplified, adapted from [1]).

In this paper, we present *Archistant*[1], an end-to-end solution for supporting the architect in conceptualizing a building (see Fig. 1 for a system overview).

---

[1] http://www.dfki.uni-kl.de/archistant.

The Archistant user interface helps the user to develop an early architectural concept. For that purpose, it is designed to follow one of the established working methods in architecture, the so-called room schedule (see Sect. 3). After such a sketch has been entered, the user can invoke the retrieval function. Archistant takes care of distributing the search query to MetisCBR and graph-matching-based based retrieval systems and to collect and unify their results. The results are sent back to the user interface, where they can be contemplated by the user. Furthermore, the user is helped to reflect the results by a mapping feature, that indicates, which room in the query relates to which room in a search result.

Until now, case-based reasoning (CBR) and graph matching have been used to retrieve the similar floor plans in separately implemented systems. The novelty of Archistant is that it takes the advantages of both methods, and combines them in one common system.

This paper is structured as follows: after the problem has been motivated and the solution roughly sketched in this section, a literature review incorporating a description of the utilized user interface is given in Sect. 2. The floor plan retrieval techniques themselves as well as the query-result mapping are stated in detail in Sect. 3. Afterwards, the system is evaluated by a stress test and qualitative evaluations of the results as well as the mapping function in Sect. 4. Finally, the paper is concluded in Sect. 5.

2   Related Work

In this section, we describe work related to our research presented in this paper. We divide this related research into three main contexts: case-based reasoning, (sub)graph matching, and sketch-based interfaces.

2.1   Case-Based Reasoning

Case-based retrieval, a sub-domain of case-based reasoning, is a technique used by previously mentioned MetisCBR to find similar floor plans. Comprehensive overviews of tools and approaches related to MetisCBR are contained in studies of Heylighen and Neuckermans [14] and Richter et al. [28]. In these two overviews, the CBR-based and related approaches were compared with different features to provide the best comparison possible for both designers (in this case architects) and academic and professional staff of the knowledge-based design domain. In [2], a table-based summary of these two studies is presented which is shown in Fig. 2. Besides this overview, we also provide descriptions of the most influential approaches that inspired the creation and development of MetisCBR.

FABEL [24] is an approach that comes very close to the current purpose of MetisCBR and has served as one of the most inspirational approaches. In FABEL, the special modules (called *specialists* in FABEL) work with cases that have a multidimensional aspect-based representation in order to find the most similar ones to a given problem (a user query which is converted to such an aspect-specific structure). The database of cases (case base) inside the FABEL

	Storage			Input			Output						
	Floor plans + text	Abstraction	Topology	Graphic	Verbal	Adaption	Reference projects	Applying solutions	Graphical Information	Learning	Subproblems	Semantic net	Analogy
Archie-II Domeshek et al. (1994)	X	X			X		X		X		X	X	
CADRE Hua et al. (1996)	X	X	X	X	X	X		X	X	X	X		
FABEL Voss (1997)	X	X	X	X	X	X	X	X	X		X		X
IDIOM Lottaz et al. (1998)			X	X	X	X		X	X				
PREC. Oxman and Oxman (1993)	X	X			X		X		X		X	X	
SEED L. Flemming (1994)		X			X	X	X	X	X		X		
SL_CB Lee et al. (2002)	X	X			X	X	X	X	X				
TRACE Mubarak (2004)		X	X	X	X		X	X	X				
CaseBook Inanc (2000)	X		X	X	X		X						
MONEO Taha (2006)	X	X			X		X		X				X
CBA Lin and Chiu (2003)	X	X			X		X				X		
DYNAMO Heylighen and Neuckermans (2001)	X	X		X	X		X		X	X		X	

Fig. 2. A tabular summary of CBR tools and approaches for architectural design support, provided in [2], of the studies by Heylighen and Neuckermans [14] and Richter et al. [28]. The comparison has three main categories: *storage, input,* and *output.* Storage is divided in *floor plans + text, abstraction,* and *topology.* Input is divided in *graphic, verbal,* and *adaption.* Output is divided in *reference projects, applying solutions, graphical information, learning, subproblems, semantic net,* and *analogy.* Figure from [2].

contains the retrievable cases where identical aspects of two cases are connected by relational arcs. The retrieval algorithm of FABEL uses a so-called *fish-and-sink* approach.

The CBR-based framework *CBArch* [8] supports the construction of buildings that have a commercial background. CBArch aims at helping the architects and other professionals involved in a construction of such a building to improve the currently developed building design by providing alternative suggestions for its configuration. CBArch considers the main architectural aspects of a building (such as size) from the energy efficiency point of view. the main functionality of CBArch supports the CBR cycle (Retrieve, Reuse, Revise, Retain). In the retrieval phase, the feature vectors are used to compare the information from query and case to assess similarity between them. The cases are also saved in a parametric ontology-based representation for graphical representation of cases.

DYNAMO (Dynamic Architectural Memory Online) is a web-based project (described in [27]) started in 1996 to provide a case base for architecture professionals and students. The service aims at providing an easy access to architectural designs by providing searching and filtering functions for the designs in the database. DYNAMO is related to other CBR approaches in the use of the dynamic memory theory of Schank [30] (in DYNAMO's case the architectural memory). Cases available in the case base of DYNAMO can be architectural designs of already existing as well as unbuilt projects, a single case consists of architectural aspects of the building as well as its graphical representation. The attribute-value-based structure underlies the representation of the cases in

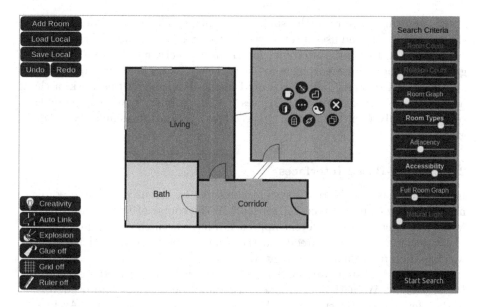

Fig. 3. Screenshot of the Archistant WebUI.

the database. The retrieval process consists of two steps: in first step, an exact matching tries to determine the structurally identical cases, after that cases from the case base are selected that have at least one criterion in common. DYNAMO can also apply Data Mining techniques, such as Collaborative Filtering.

One of the first CBR-related architectural design support applications is *CADRE* (description of which is available in [29]), developed between 1990–1994. CADRE was constructed to work with 3D models of buildings and extends the model with some features that can emphasize its context (e.g., the environmental criteria such as street context or direction, or topological features such as room transformation). However, CADRE does not implement a retrieval component (a user her/himself should select a proper case from the case base) and concentrates on adaptation of solutions, i.e., the Reuse phase of the CBR cycle. In this phase, CADRE tries to adapt the existing solution into a new environment with given constraints. A successor of CADRE is the *IDIOM* system that instead of using the 3D models concentrates on the 2D-based representations of floor plans (or parts of them).

2.2   Graph and Subgraph Matching

Graph matching is widely applicable nowadays for its usability in matching and retrieval problems. In real life scenarios, there are situations when there is no exact match with the whole graph but there is a part (subgraph) that matches. If a subgraph is available, then we can use subgraph matching that tells us about the parts of two graphs that are isomorphic. Technique presented in [2] used graph matching to retrieve the similar floor plans. This work slightly modifies the method in [19] and uses it for retrieval of similar floor plans by arranging the row-column vectors of the adjacency matrix in the decision tree. Work in [25] uses graph and subgraph isomorphism to check the similarity between a query graph and models of different buildings stored in the database. In this work, a check has also been implemented that ensures that if the query graph corresponds to some rules only then the system proceeds for checking the similarity of query graph.

2.3   Sketch-Based Interfaces

In order to make the retrieval system accessible by the user, a sketch editor is needed to enter an architectural concept. In the course of the research project Metis (see Sect. 5), two different approaches have been compared [6]. The first (Touchtect) was based on free-hand sketching, while the other (Metis WebUI) was based on polygonal rooms (Fig. 3).

The retrieval system presented here is accessible by a dedicated user interface, the Archistant WebUI. This browser-based application is an improved version of the Metis WebUI (first described in [6]). The main purpose of the Archistant WebUI is to help the user develop an architectural concept and thus generate retrieval queries. The general usability of the Metis WebUI has been shown by the means of a user study. For query construction, the Archistant WebUI uses the AGraphML [16] specification (also see Sect. 3.1).

The Archistant WebUI provides a room-oriented floor plan editor. Is designed to follow the room schedule working method as established in architecture. A room-schedule in architecture is a set of high-level requirements (usually coming from an end-customer or contractor), that has to be turned into a floor plan layout by the architect. Its formal structure is assumed to be a graph in the course of this paper. Hence, attributes of rooms are modeled as node attributes and attributes of room connections are modeled as edge attributes. Rooms are created in an abstract, shapeless mode indicated by a circle. Rooms may always be dragged independently from each other and their attributes and connections can iteratively refined by the user, where each aspect can be specified as abstract or specific as desired. As a convention, a single line wall between two rooms indicate a wall connection, double lines represent doors.

In order to be usable as a search interface, the Archistant WebUI possesses a search sidebar, in which the fingerprint weights can be adjusted and the retrieval process can be triggered. Furthermore, the result thumbnails are also shown here as well as the full screen query-result mapping view (showing which room of the query relates to which room of the result) can be invoked. Finally, results can be rated by the user, allowing for machine learning-based optimization in future.

3   Floor Plans Retrieval Techniques

In this Section, we present the main components and underlying concepts for our floor plans retrieval framework that combines three different search methods for this purpose. The framework is an integral part of the Archistant infrastructure and allows for comprehensive search process with CBR and two (sub)graph matching methods: VF2 (exact matching) and IB (index-based retrieval based on the Neo4j graph database index). The underlying structure for search of similar floor plans is the paradigm of semantic fingerprint that allows for decomposition of the search request into different semantically enhanced sub-patterns, thus giving us opportunity to look for the best fit for the floor plan query based on the features that are important for this particular query only. In Sect. 3.1 we describe the *semantic fingerprint*, i.e., our underlying sub-patterns paradigm, followed by Sect. 3.2 that briefly describes our query structure. In Sect. 3.3 we present our three retrieval methods, including CBR-based MetisCBR and (sub)graph-based VF2 and IB.

3.1 Semantic Fingerprints Concept

Langenhan and Petzold [17] describe semantic fingerprint as a hierarchically constructed index for definition of floor plans that enhances the well-known concept of Building Information Modeling (BIM). To represent the fingerprints, a graph-based structure is developed that can represent the topology of the floor plan and the connections between particular node units (rooms) including only the graph attributes defined for this fingerprint. To transform this graph-based structure into a machine-readable format (XML), the AGraphML specification [16] is used. Furthermore, semantic fingerprint is a representative of room-based configuration, thus rooms and their relations play the most important role in resolving queries that are constructed in the same way. Our searching techniques can detect a number of fingerprints in the query provided by the user: VF2 and IB apply the decomposition of the floor plan query, whereas MetisCBR implements the recognition of patterns based on the fingerprints data contained in the query. In Fig. 4 a list of 7 fingerprint patterns that are common for each of our searching techniques are shown.

3.2 Query Structure

Retrieval queries in Archistant are constructed by utilizing the AGraphML structure: for each room in the floor plan concept, a node is created and the room's properties are used as node properties. Likewise, connections between rooms are represented as edges and the connection's attributes are used as edge attributes. Finally, the resulting AGraphML is wrapped into a search query XML structure along with the user-defined fingerprint weights. In Listing 1.1, a general structure for our queries is shown.

Fingerprint	Name	Description	Specifics
FP1 	Room Count	Number of rooms	No connections between rooms and no labels specified
FP2 	Relation Count	Number of edges	No room information specified
FP3 	Room Graph	Anonymous representation of rooms and edges	No labels specified for rooms and edges
FP4 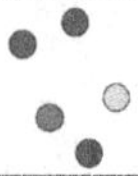	Room Types	Labels of rooms	No connections between rooms only room labels are specified
FP5 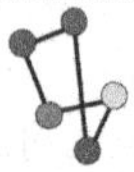	Adjacency	Emphasis on room semantics	Rooms information is complete no edge labels specified
FP6 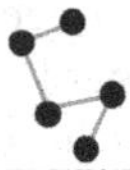	Accessibility	Emphasis on edge semantics	Edge information is complete no room labels specified
FP7 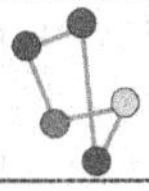	Full Graph	Complete graph	All information about rooms and edges available

Fig. 4. Fingerprint patterns currently available in all three (MetisCBR, VF2, IB) retrieval techniques of Archistant (Figure from [1]).

Listing 1.1. General structure of a query for the retrieval methods in our framework (adapted version from [5]).

```
<?xml version="1.0" encoding="UTF-8"?>
<searchrequest>
        <fingerprint name="Room_Types" weight="0.7" />
        <fingerprint name="Adjacency" weight="0.3" />
        <agraphml>
                <graphml>...</graphml>
        </agraphml>
</searchrequest>
```

3.3   Matching Techniques

Case-Based Retrieval (MetisCBR). MetisCBR was developed to apply a multi-agent system with case-based agents to problems of retrieval of similar floor plans during the early phases in architectural conceptual design. Its main features are the retrieval containers that can concurrently resolve different queries that may belong to the same retrieval process (or be completely independent, i.e., triggered by another retrieval process). Before the actual retrieval takes places, the search request is analyzed, divided in the sub-queries (if multiple semantic fingerprints were detected in the request), and then assigned to the corresponding retrieval container that consists of the agents most suitable for this type of query/fingerprint. This assignment process is governed by a special coordinator agent described in [5]. Figure 5 shows a general overview of the MetisCBR system.

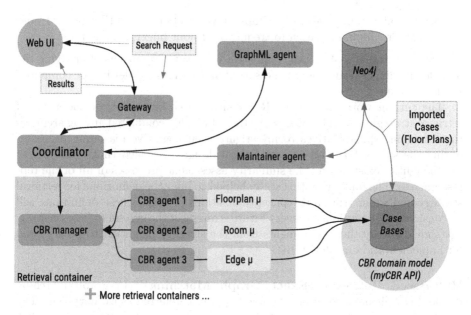

Fig. 5. General architecture of MetisCBR (a detailed description is available in [4,5]).

As a CBR-based system, MetisCBR defines an underlying structure for each case saved in its case base. This structure is mostly based on a domain model. For MetisCBR, a distributed domain model (described in [3]) was created to govern the system's cases. Each case represents a single floor plan and is divided into three main concepts: FLOORPLAN (meta data about the floor plan), ROOM (information about rooms), and EDGE (information about edges, i.e. room connections). Attributes, such as `roomType` or `windowExist` for rooms, and `edgeType` and `linearDistance` for edges define the detailed structure of a case.

The attributes are combined in different amalgamation functions that either correspond to semantic fingerprints or can be of generic type. It depends on actions of the user (who may or may not include the fingerprint patterns in the query) which amalgamation will be used for the current search. For the amalgamations that are connected to the fingerprints, a combination of attributes is selected for the search that is predefined and unique for this fingerprint only (it is of course possible that an attribute is available in multiple fingerprints, i.e., an attribute can be used multiple times during the same search process). In [22] a footprint sets based retrieval system is presented that became an inspiration for our fingerprint-amalgamation-based retrieval. The fingerprint amalgamation and the generic amalgamation (that uses all attributes for comparison) can be used in two different types of retrieval strategies:

- A strategy for fingerprints that have a complicated structure (such as FP5 or FP6) and a comprehensive search without fingerprints defined. This strategy is presented in [3].
- Faster strategy that uses more simple fingerprints (such as FP1 or FP4) and applied for simplified search for requests without fingerprints defined.

After the actual search, the results can be elevated by means of applying the user-defined fingerprint weights and sorted in descending order by the computed similarity value.

The current work on MetisCBR is concentrated on further development of retrieval strategies. A study of Ayzenshtadt et al. [26], conducted among architectural domain representatives to investigate their cognitive reasoning processes during the search for similar architectural designs, revealed that a number of commonalities exist among the similarity assessment processes of all of the representatives. The findings of this study helped to infer the definitions for retrieval strategy and superstructural (conceptualization) process. These definitions will be considered foundations for every future strategy of the system (e.g., each strategy should satisfy the requirements from the strategy definition to be accepted for implementation in MetisCBR).

VF2-Based Retrieval (Exact Graph Matching). In graph matching domain, the phenomena of one-to-one mapping is referred as isomorphism. The graphs are isomorphic when they follow the exactly same topology, that is, they both have the same number of nodes and each of the corresponding node is connected in same way. Exact graph matching is a way to detect the isomorphism [7]. Some of the one-to-one exact graph-based matching approaches include: [18,20,23]. For Archistant, we decided to use the VF2 algorithm, proposed in [11], its implementation is provided in the NetworkX library. As compared to other available implementations, VF2 has the capacity to achieve the best performance for small and sparse graphs [12]. In addition to this, it requires less memory.

Our exact graph matching system (VF2) relies on a preprocessing step. During preprocessing, one AGraphML file is generated for each of the floor plans in the data base. Later on, these AGraphML files are used by the VF2 system. A tool named "Neo4j Shell Tools" is available that is used to generate these AGraphML files.

VF2 system performs different steps in order to compare the search request with the floor plans in the data base (see Fig. 6). Firstly, once the request is received by the system, its validity is ensured. Only the valid requests are forwarded to the next step. In this second step, AGraphML is extracted from

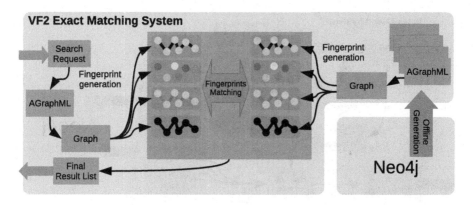

Fig. 6. The above diagram shows the workflow of the VF2 exact matching system. It shows the step by step details of how a search request and the floor plans in data base are decomposed into fingerprints and then their corresponding fingerprints are matched. Finally, the results are transferred to the requester (Figure from [1]).

the search request to generate a graph, that is referred as *query-graph*. The *query-graph* represents nodes and connections between the nodes. Finally, the *query-graph* is decomposed into fingerprints. All the aforementioned steps take place each time, when the user creates a query, before the actual matching part. Once the fingerprints are generated for the *query-graph*, the fingerprints for floor plans in the data base need to be generated. For this purpose VF2 system, one by one takes each of the AGraphML files, referred as *db-graph*, and generates its fingerprints. These fingerprints are then matched with the fingerprints of the *query-graph*. Each of the corresponding fingerprints, that is, FP1 of *query-graph* is matched against FP1 of *db-graph*, FP2 of *query-graph* is matched against FP2 of *db-graph* and so on. Based upon the matching fingerprints, a similarity score is computed, that shows how closely a *db-graph* is similar to the *query-graph* (see Fig. 4). Finally, VF2 system sends back the results with top similarity scores in descending order.

Index-Based Retrieval (Inexact Graph Matching). Several different approaches of index-based graph matching methods have been described in literature, including GraphGrep [13], Lucene index [21], FG-Index [10] and cIndex [9]. Archistant's index-based retrieval uses Lucene index since this indexing method is used by the Neo4j database by default.

The index-based retrieval can be described as follows (see Fig. 7): A search request AGraphML file is decomposed into the different fingerprints and a set of fingerprint weights. Graph-based fingerprints are represented by internal graph structures. These internal representations are then translated into cypher queries and successively passed to the Neo4j server.

The Neo4j server replies each request with a set of floor plans (more precisely URIs referencing the floor plans are used). These sets are unified discarding all

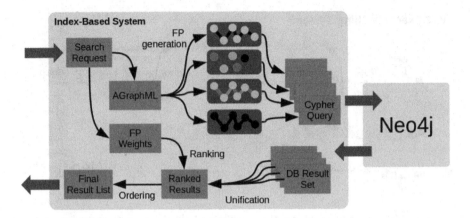

Fig. 7. The above figure shows index-based retrieval flow chart. After the search request is decomposed into a set of internal graph structures representing the different finger-prints, cypher queries for each fingerprint are created. These cypher queries are successively passed to the Neo4j server, and the Neo4j's replies to each query with a set of floor plan references. These references are unified, taking the user-defined weighting into account (Figure from [1]).

redundant entries. Simultaneously, the index-based similarity score is calculated for every item. This similarity score is the sum of the user-defined weights of the fingerprints for which the query matches the database entry. Finally, the result list is brought into descending order according to the index-based similarity score.

A graph-based fingerprint is considered to match a database entry if the fingerprint's graph is a subgraph of the database entry. The fingerprints are processed independently from each other for simplicity reasons, hence one room in the query may be mapped to different rooms within the same floor plan in the database. Figure 8 illustrates an example of the fingerprints processing within the index-based method. The query consists of three rooms labeled as Living, Kitchen and Sleeping. The Living room is connected with Kitchen via an edge connection labeled as Passage, the Kitchen is connected with Sleeping via an edge connection labeled as Wall, and Sleeping room is connected with Living via an edge connection labeled as Door. The right side of the diagram shows exemplary matching and non matching fingerprints between search query and floor plan in the database.

3.4   Augmentation of Retrieved Floor Plans

The retrieval systems deliver results in the form of URLs which point to plain image files. These image files serve as thumbnails for the individual results. In order to allow for better user experience, additional information is needed: firstly, detailed information about the results' graph structures allows for rendering of the results' floor plans in higher quality. By using the same layout as employed

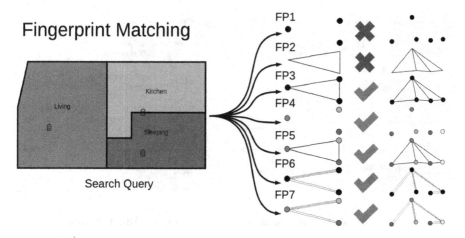

Fig. 8. Decomposition of a floor plan query into fingerprints and subsequent matching with a database entry (Figure from [1]). Depicted is the subgraph matching behavior as implemented in the index-based retrieval.

in the WebUI editor, the user may orient himself more easily in the results. Secondly, a map from individual rooms in the user's query to individual rooms in the server's results helps the user to understand how the results have been derived. These informations are gathered and centralized for all results of all retrieval systems at the augmentation processor (AP, see Fig. 9).

Generation of Result-Related AGraphML Files. Both the generation of AGRaphML files related to the result image file URLs and the generation of room maps from query rooms to result rooms are implemented by querying the same Neo4j database on which the retrieval systems are based. As a basic principle, the image URLs used by the retrieval systems are also attached to the graphs in the database. The generation of result AGraphML files is implemented as follows:

1. Result image URL is used to retrieve the id of a so-called storey vertex. These storey vertices are used to organize floor plans. All room-representing nodes of a floor plan are connected to a single storey vertex.
2. The node IDs (along with the relevant node properties like room purpose and room layout polygon) of all room nodes connected to the storey vertex of interest are obtained.
3. All connections between the nodes retrieved earlier are obtained.

Generation of Room Maps. The AP uses Neo4j's matching mechanisms to obtain maps from the user's query to the retrieval system's results. Therefore, the fingerprint abstractions are employed here just like in the retrieval systems.

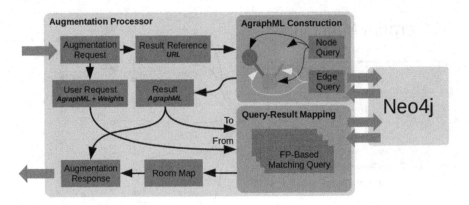

Fig. 9. Structure of the augmentation processor (Figure Adapted from [31]).

Based on the order of the user's fingerprint weights, different abstractions of result and query are tried to be matched. Since not all retrieval systems use exact matching techniques (and not all fingerprints are selected as mandatory by the user), there might be results to which no fingerprints of the query match the result's fingerprints at all. The first matching fingerprint (where the order is determined by the user) is used for the generation of the final room map. There are situations, in which the abstractions of result and query match in more than one way (e.g. in FP1, any room may match). In such cases, one of the matches with the highest number of matching room purposes is selected (if there are multiple of them, randomly). A visualization of the room map can be displayed to the user (see Fig. 10).

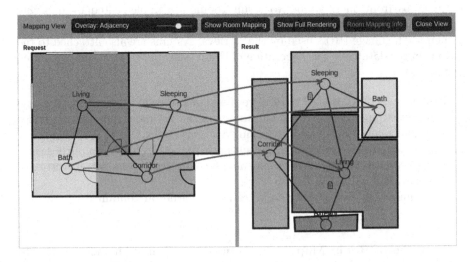

Fig. 10. Screenshot of the room mapping view in the Archistant WebUI.

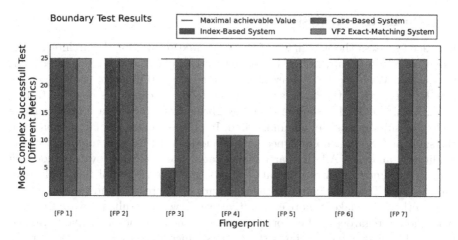

Fig. 11. Overview of the boundary test results. For each fingerprint and retrieval system the boundary is depicted (the metrics differ between different fingerprints). For each fingerprint, the maximum achievable value is given (Figure from [1]).

4   Evaluation of Our System

4.1   Computational Limitations (Boundary Test)

All the presented algorithms are expected to terminate properly for any given search query in theory. However, since both graph matching and CBR are computational demanding, there are practical limits (boundaries) to the complexity of a search query our system can handle. In order to determine these boundaries, we conducted an automated stress-test in which for every fingerprint we run a series of test cases and record the behavior of the retrieval systems. In each series, test cases of increasing complexity are used. In most cases, a test scenario is considered of complexity n, if it consists of n rooms. For FP2 however, n connections are used instead. For graph-based fingerprints, we use linear graphs. Based on the type of fingerprint, we used different node and edge attributes, that are randomly selected for each test case. A boundary of a retrieval system for a certain fingerprint is considered to be the complexity rating of the lowest test case if the system was unable to process without crash minus 1. The results of the boundary test we conducted are depicted in Fig. 11. Both the VF2-based retrieval and the case-based retrieval managed to process all test cases without crash. Only the index-based retrieval system exhibited limitations over the maximum size of fingerprints FP3, FP5, FP6 and FP7. It is assumed that these limitations arose from internal timeout errors of the underlying Neo4j database. Generally, given more memory and computational power, these boundaries could be raised. Using the determined boundaries of the different retrieval systems, the productive use of Archistant can be secured by restricting the system to queries that are known to be manageable.

4.2   Qualitative Analysis

In order to assess the usefulness of the generated results for the architects, the quality of results is subjectively estimated for a set of dedicatedly generated sample queries. With the help of architects, we created 10 different search queries. One by one, each of the queries was entered to the Archistant's retrieval framework. Archistant performed the same fingerprint matching for each of the queries and the results were observed and stored. In order to make the fair comparison, the same architects, who designed search queries with us, also took part in the qualitative analysis. All the participants rated the three retrieval methods on scale of 1st, 2nd, and 3rd or equal. Table 1 shows the results of this qualitative rating study. The table contains the summary of the results and the ratings for each method. To make it more elaborative, we show the results in two categories. A method is regarded as best or clear winner when all the participants ranked it best. The first category shows the queries which were ranked as best for the corresponding retrieval method. The second category shows the results of methods that were considered as best by majority of the participants for a particular query. The third column shows the queries that got equal number of votes. The last column shows the percentage of the dominating queries.

It is clear from Table 1, that none of the retrieval methods failed completely, rather they were able to produce quality results of some of the queries. Randomly, we selected three queries, their two best results for each of the retrieval methods are shown in Fig. 15. For ease of understanding we show the results in graph-based representation and graphical representation shown by top and bottom representation in Table 1 respectively. It is worth noting that the CBR method has a higher score in relation to other two methods.

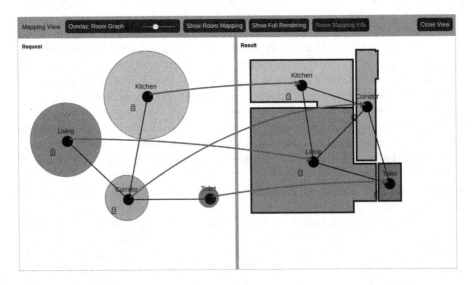

Fig. 12. Mapping between query and result 1 (room graph fingerprint overlay).

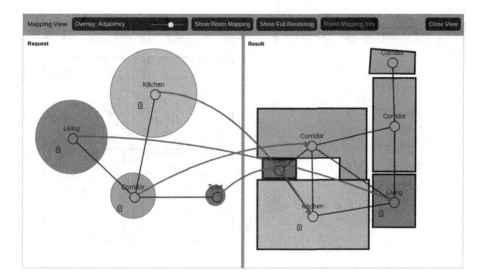

Fig. 13. Mapping between query and result 1 (adjacency fingerprint overlay).

In general, we can notice that the retrieval methods were able to find and present sets of reasonable results. Subjectively, the VF2 method outperformed other techniques overall, confirming the assumption that the exact isomorphism can be seen as the most suitable method for matching in databases with certain structural and technical constraints.

4.3   Query-Result Mapping Case Study

In order to demonstrate the usefulness of the query-result mapping functionality, we applied this algorithm to a query selected from the qualitative study described above.

Given was a sketch with a living room, a kitchen, a toilet, and a corridor, where all rooms are connected to the corridor by passage connections. The retrieval system returned several results, from which 3 were investigated. In Fig. 12, the query is matched to a floor plan that has exactly the same amount of rooms, all room functions in the query are matched to rooms with same functions. However, the architect might at least get inspiration for a room layout. In the second retrieval result (see Fig. 13), the queried structure is mapped to a larger one, that could inspire the architect to make some additions to his concept. Finally, in the third result (Fig. 14), a graph structure is found, that is also extended compared to the query. When switching to the full room graph overlay, it became obvious, that all room functions could be matched, not all connections could. However, the founded mapping is suggesting new connection types. This could hint the user that a doorless connection between a corridor and a toilet may be improved.

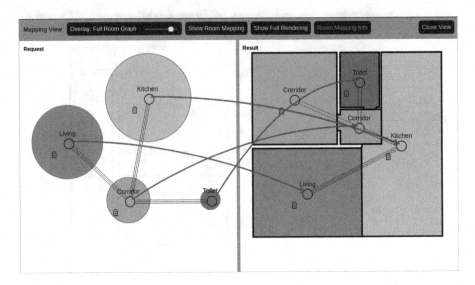

Fig. 14. Mapping between query and result 1 (full room graph fingerprint overlay).

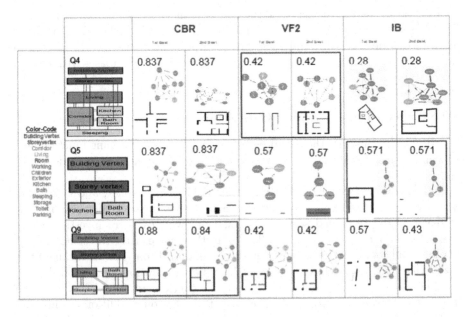

Fig. 15. The above figure shows the similarity scores of selected queries. Color codes represent the room purposes, the first column contains the queries with rooms and assigned purposes. The two best results of a query against each retrieval method are shown. Each box shows the similarity score, the corresponding graph, and a graphical floor plan representation. The colored boxes show the best results (Figure from [1]). (Color figure online)

Table 1. The above table shows the results of selected queries for each of the retrieval method. The second column shows the queries whose results were ranked as best by all the participants against the corresponding retrieval method in the first column. The third column shows the queries which won the support of majority of participants. Queries that get equal number of votes are placed in fourth column. The last column contains the percentage of queries dominated by the particular method (Table from [1].)

Retrieval method	Queries won	Queries won by majority	Co-winner in	Summarized results
VF2	Q3, Q4	Q6, Q7	Q8, Q10	50%
Index-Based	–	Q5	Q10	15%
MetisCBR	Q2, Q9	Q1	Q8	35%

5 Conclusion and Future Work

In this work, we presented a novel possibility for architects to enhance the early conceptual design phase by using an end-to-end system Archistant that is able to search for similar floor plans during this phase of the design process. Archistant uses a sketch-based interface for construction of floor plan queries and distributes this query, with the help of a processing component, among three different retrieval methods that are based on different research paradigms of artificial intelligence, namely, case-based reasoning and graph matching. The retrieval results are enhanced by an augmentation processor that is able to visualize room mapping between the query and the corresponding result, thus providing a justification of how both room configurations match. We evaluated the complete system with a boundary test to determine the retrieval-related limitations of our three searching techniques, where most of the methods were able to deal with the highest complexity of a query. We also conducted a qualitative analysis where each of the retrieval methods was able to satisfy the expectations on the delivered results in at least some of the cases. Our future work will concentrate on building of a bigger collection of retrievable floor plans, and an inclusion of machine learning for automatic improvement of results.

Acknowledgement. *MetisCBR* [4] and the *Metis WebUI* have been implemented in the course of the research project *Metis – Knowledge-based search and query methods for the development of semantic information models (BIM) for use in early design phases*. Metis is an interdisciplinary project, funded by the German Research Foundation (Deutsche Forschungsgemeinschaft, DFG).

References

1. Sabri, Q.U., Bayer, J., Ayzenshtadt, V., Bukhari, S.S., Althoff, K.-D., Dengel, A.: Semantic pattern-based retrieval of architectural floor plans with case-based and graph-based searching techniques and their evaluation and visualization. In: De Marsico, M., di Baja, G.S., Fred, A.L.N. (eds.) Proceedings of the 6th International Conference on Pattern Recognition Applications and Methods, (ICPRAM 2017), 24–26 February, Porto, Portugal, pp. 50–60. SCITEPRESS (2017). ISBN 978-989-758-222-6

2. Ahmed, S., Weber, M., Liwicki, M., Langenhan, C., Dengel, A., Petzold, F.: Automatic analysis and sketch-based retrieval of architectural floor plans. Pattern Recogn. Lett. **35**, 91–100 (2014)

3. Ayzenshtadt, V., Langenhan, C., Bukhari, S.S., Althoff, K.-D., Petzold, F., Dengel, A.: Distributed domain model for the case-based retrieval of architectural building designs. In: Petridis, M., Roth-Berghofer, T., Wiratunga, N., (eds.) Proceedings of the 20th UK Workshop on Case-Based Reasoning, (UKCBR 2015), located at SGAI International Conference on Artificial Intelligence, 15–17 December, Cambridge, United Kingdom. School of Computing, Engineering and Mathematics, University of Brighton, UK (2015)

4. Ayzenshtadt, V., Langenhan, C., Bukhari, S.S., Althoff, K.-D., Petzold, F., Dengel, A.: Thinking with containers: a multi-agent retrieval approach for the case-based semantic search of architectural designs. In: Filipe, J., van den Herik, J. (eds.) Proceedings of the 8th International Conference on Agents and Artificial Intelligence, (ICAART 2016), 24–26 February Rome, Italy. SCITEPRESS (2016)

5. Ayzenshtadt, V., Langenhan, C., Roith, J., Bukhari, S., Althoff, K.-D., Petzold, F., Dengel, A.: Comparative evaluation of rule-based and case-based retrieval coordination for search of architectural building designs. In: Goel, A., Díaz-Agudo, M.B., Roth-Berghofer, T. (eds.) ICCBR 2016. LNCS (LNAI), vol. 9969, pp. 16–31. Springer, Cham (2016). https://doi.org/10.1007/978-3-319-47096-2_2

6. Bayer, J., Bukhari, S.S., Langenhan, C., Liwicki, M., Althoff, K.-D., Petzold, F., Dengel, A.: Migrating the classical pen-and-paper based conceptual sketching of architecture plans towards computer tools - prototype design and evaluation. In: Lamiroy, B., Dueire Lins, R. (eds.) GREC 2015. LNCS, vol. 9657, pp. 47–59. Springer, Cham (2017). https://doi.org/10.1007/978-3-319-52159-6_4

7. Bengoetxea, E.: Inexact graph matching using estimation of distribution algorithms. Ecole Nationale Supérieure des Télécommunications, Paris **2**(4) (2002)

8. Cavieres, A., Bhatia, U., Joshi, P., Zhao, F., Ram, A.: CBArch: a case-based reasoning framework for conceptual design of commercial buildings. In: Artificial Intelligence and Sustainable Design - Papers from the AAAI 2011 Spring Symposium (SS-11-02), pp. 19–25 (2011)

9. Chen, C., Yan, X., Yu, P.S., Han, J., Zhang, D.-Q., Gu, X.: Towards graph containment search and indexing. In: Proceedings of the 33rd International Conference on Very Large Data Bases, pp. 926–937. VLDB Endowment (2007)

10. Cheng, J., Ke, Y., Ng, W., Lu, A.: FG-Index: towards verification-free query processing on graph databases. In: Proceedings of the 2007 ACM SIGMOD International Conference on Management of Data, pp. 857–872. ACM (2007)

11. Cordella, L.P., Foggia, P., Sansone, C., Vento, M.: A (sub) graph isomorphism algorithm for matching large graphs. IEEE Trans. Pattern Anal. Mach. Intell. **26**(10), 1367–1372 (2004)

12. Foggia, P., Sansone, C., Vento, M.: A performance comparison of five algorithms for graph isomorphism. In: Proceedings of the 3rd IAPR TC 2015 Workshop on Graph-based Representations in Pattern Recognition, pp. 188–199 (2001)

13. Giugno, R., Shasha, D.: GraphGrep: a fast and universal method for querying graphs. In: Proceedings of the 16th International Conference on Pattern Recognition, vol. 2, pp. 112–115. IEEE (2002)

14. Heylighen, A., Neuckermans, H.: A case base of case-based design tools for architecture. Comput.-Aided Des. **33**(14), 1111–1122 (2001)

15. Inanc, B.S.: Casebook. An information retrieval system for housing floor plans. In: The Proceedings of 5th Conference on Computer Aided Architectural Design Research (CAADRIA), pp. 389–398 (2000)

16. Langenhan, C.: A federated information system for the support of topological bim-based approaches. In: Forum Bauinformatik Aachen (2015)

17. Langenhan, C., Petzold, F.: The fingerprint of architecture-sketch-based design methods for researching building layouts through the semantic fingerprinting of floor plans. Int. Electron. Sci.-Educ. J.: Archit. Mod. Inf. Tech. **4**, 13 (2010)

18. McKay, B.D., et al.: Practical graph isomorphism. Department of Computer Science, Vanderbilt University Tennessee, US (1981)

19. Messmer, B.T., Bunke, H.: A decision tree approach to graph and subgraph isomorphism detection. Pattern Recogn. **32**(12), 1979–1998 (1999)

20. Schmidt, D.C., Druffel, L.E.: A fast backtracking algorithm to test directed graphs for isomorphism using distance matrices. J. ACM (JACM) **23**(3), 433–445 (1976)

21. Sharanya Jayaraman, S.V.: Comparative survey of query processing on graph databases. Project report, Florida State University (2013)

22. Smyt, B., McKenna, E.: Footprint-based retrieval. In: Althoff, K.-D., Bergmann, R., Branting, L.K. (eds.) ICCBR 1999. LNCS, vol. 1650, pp. 343–357. Springer, Heidelberg (1999). https://doi.org/10.1007/3-540-48508-2_25

23. Ullmann, J.R.: An algorithm for subgraph isomorphism. J. ACM (JACM) **23**(1), 31–42 (1976)

24. Voss, A.: Case design specialists in FABEL. In: Issues and Applications of Case-Based Reasoning in Design, pp. 301–335 (1997)

25. Wessel, R., Blümel, I., Klein, R.: The room connectivity graph: shape retrieval in the architectural domain (2008)

26. Ayzenshtadt, V., Langenhan, C., Bukhari, S., Althoff, K.-D., Petzold, F., Dengel, A.: Extending the flexibility of case-based design support tools: a use case in the architectural domain. In: Aha, D.W., Lieber, J. (eds.) ICCBR 2017. LNCS (LNAI), vol. 10339, pp. 46–60. Springer, Cham (2017). https://doi.org/10.1007/978-3-319-61030-6_4

27. Heylighen, A., Schreurs, J., Neuckermans, H.: ENTER instead of SUBMIT. In: DesignNet-Knowledge e Information Management per il design, pp. 171–182 (2002)

28. Richter, K., Heylighen, A., Donath, D.: Looking back to the future-an updated case base of case-based design tools for architecture. Knowl. Model.-eCAADe **25**, 285–292 (2007)

29. Richter, K.: Augmenting designers' memory: case based reasoning in der Architektur. Logos-Verlag (2011). ISBN 9783832527334

30. Schank, R.C.: Dynamic Memory: A Theory of Reminding and Learning in Computers and People. Cambridge University Press, Cambridge (1983)

31. Bayer, J.: Development of a modular software framework for supporting architects during early design phases. Master's thesis, University of Kaiserslautern (2017)

Characterization of a Virtual Glove
for Hand Rehabilitation Based
on Orthogonal LEAP Controllers

Giuseppe Placidi[1]([✉]), Luigi Cinque[2], Matteo Polsinelli[1],
and Matteo Spezialetti[1]

[1] A²VI_Lab, Department of Life, Health and Environmental Sciences,
University of L'Aquila, Via Vetoio, L'Aquila, Italy
giuseppe.placidi@univaq.it, matteo.polsinelli@student.univaq.it,
matteo.spezialetti@graduate.univaq.it
[2] Department of Computer Science, Sapienza University, Via Salaria, Rome, Italy
cinque@di.uniroma1.it

Abstract. Hand rehabilitation therapy is fundamental for post-stroke
or post-surgery impairments. Traditional rehabilitation requires the presence of a therapist for executing and controlling therapy: this implies high
costs, stress for the patient, and subjective evaluation of the therapy
effectiveness. Alternative approaches, based on mechanical and tracking-
based gloves, have been recently proposed. Mechanical devices are often
expensive, cumbersome and patient specific, while tracking-based devices
are not subject to this limitations, but, especially if based on a single
tracking sensor, could suffer from occlusions. In this paper a multi-sensors
approach, the Virtual Glove (VG), based on the simultaneous use of two
orthogonal LEAP motion controllers, was presented. In particular, the
VG design was summarized, an engineered version was presented and its
characterization was performed through spatial measurements. Measurements have been compared with those collected with a accurate spatial
positioning system for evaluating the VG precision. The proposed strategy described the procedure to be used for VG assembly and for making
it to correctly operate.

Keywords: Hand rehabilitation · Virtual Glove
LEAP motion controller

1 Introduction

For patients suffering from post-stroke or post-surgery residual impairments,
the recovery of the hand functions is extremely important for accelerating the
rehabilitation process and it depends on frequency, duration and quality of the
rehabilitation sessions [1–4]. Traditional rehabilitation requires that a therapist
follows the patient during long and challenging one-to-one sessions. Moreover, the
effectiveness of the procedure is evaluated subjectively by the therapist, basing on

© Springer International Publishing AG, part of Springer Nature 2018
M. De Marsico et al. (Eds.): ICPRAM 2017, LNCS 10857, pp. 190–203, 2018.
https://doi.org/10.1007/978-3-319-93647-5_11

his experience. Over the last years, several automated (tele)rehabilitation gloves, based on mechanic devices or tracking sensors have been presented, for allowing patients to execute therapy in a domestic environment, while its effectiveness is numerically evaluates and controlled by therapists through Internet [5–13].

Mechanical gloves (MG) are equipped with pressure sensors and pneumatic actuators for assisting and monitoring the hand movements and to apply forces to which the fingers have to oppose [14,15]. MG are expensive, cumbersome and patient specific (reusing for other patients or for the other hand of the same patient is impossible). Tracking-based gloves consist on computer vision algorithms for the analysis and interpretation of videos from depth sensing sensors to calculate hand kinematic information in real time, [13,16–21]. Besides depth sensors, the LEAP motion controller [22] is a small and low-cost hand 3D tracking device characterized by high-resolution and high-reactivity [23], represents a good system to be used for virtual reality applications [24] and has been recently also used for the hand rehabilitation [13].

The rehabilitation system proposed in [13] consisted on two orthogonal LEAPs used for reducing the occurrence of occlusions. The two LEAPs have been fixed to a support that maintain them orthogonal each other, each at a distance of 25 cm from the corner of the support, for creating a wide area in which the hand can freely move and tracked by both sensors. In this paper we present an engineered version of the LEAP based Virtual Glove and a characterization of the proposed system by using a numerically controlled machine in order to allow an accurate system calibration and positioning error quantification. Experimental measurements are reported and discussed.

The remaining of the paper is organized as follows: Sect. 2 summarizes the system set up and the used tracking strategy; Sect. 3 describes the system calibration; Sect. 4 presents the system characterization, by using an accurate spatial positioning system, discusses the positioning errors and presents some preliminary hand tracking results; Sect. 5 concludes the paper and describes future developments and applications.

2   System Set Up

The VG system is designed to obtain simultaneous information from two LEAPs (Fig. 1a), orthogonally placed each other (Fig. 1b). In fact, a single LEAP sensor is unable to compute with accuracy the hand position if the palm (and the fingers) is not visible, when the hand is approximately orthogonal to the sensor plane. Using two orthogonal sensors should ensure that at least one of them is able to get the correct position. The support used for lodging the two LEAPs was realized in aluminium and the lodges have been carved through a numerical control mill (GUALDONI Mod. FU 80, 1995, Milan (Italy), spatial precision 0.01 mm) of the same shape and dimension of a LEAP cover (see Fig. 2a for the support details). The LEAPs were fixed in position inside the lodges through plastic screws, just for avoiding possible vibrations and movements (see Fig. 2b for the assembly details). The centre of each LEAP was positioned at 18.5 cm

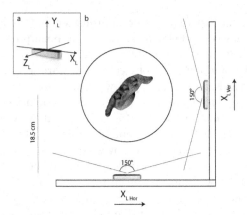

Fig. 1. (a) The LEAP sensor and its references system. (b) Hardware configuration with two orthogonal LEAPs was designed for creating a sufficiently wide area in which the hand is tracked (image from [13]).

from the internal part of the corner of the support (in the first prototype it was 25 cm): this was established for maximizing the signal in a cylindrical region of interest (ROI) of radious 10 cm and height 22 cm (the cylinder axis was perpendicular to the plane of the VG support), while also reducing the VG dimensions. In fact, by using the experience gathered in [13], the effects of direct infrared radiation from one LEAP to the other was negligible and the positioning precision worsen with the distance from the sensors: for these reasons, the minimum distance for improving signal quality into he desired useful ROI was chosen. The design and construction of the support was very accurate with respect to the first prototype presented in [13] in order to ensure that a rotation of 90° around the axis perpendicular to the plane of the support was the only movement necessary for overlapping one LEAP to the other.

One of the major issues to be addressed was the devices connection: it is impossible to manage multiple instances of the LEAP on the same machine. For this reason an architecture including a virtual machine was designed. The virtual machine (Slave) was installed on the physical machine (Master): in this way, plugging both sensors, one of them was assigned to the Master and the other to the Slave for allowing to the machines to instantiate their own driver. On each machine, data provided were captured and rerouted towards a server (hosted on the Master machine). In this way, the server was able to send data of both devices to one or more clients running on the Master.

The hand tracking system used for the proposed VG is based on a mutual exclusion strategy. The algorithm used for obtaining the positions, illustrated in Fig. 3, is based on a control switching approach: both LEAPs acquired their frames stream, but only one, depending on the rotation of the hand with respect to the horizontal LEAP references system, is used to represent the hand (Fig. 4). Both LEAPs are simultaneously turned on and remain in this state for the whole session because the time necessary to switch on and off a sensor would be

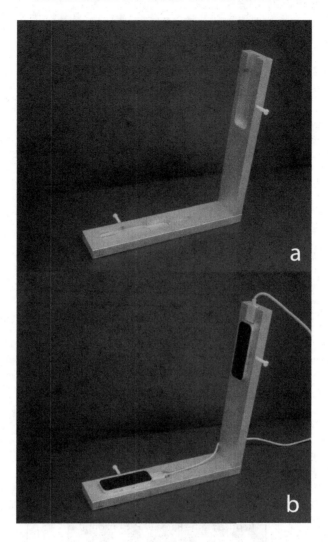

Fig. 2. (a) VG support and (b) the mounted VG system.

impossible in real-time. At each time, only one LEAP is selected as "active" and the corresponding frame is acquired and used to track the hand. The vector of hand palm **v**, orthogonal to the palm and used by the sensor software to estimate the hand orientation, is used for computing the roll r of the hand, that is the angle between the x axis and the projection of the vector on the $x-y$ plane, with respect to the horizontal LEAP reference system. As shown in Fig. 4, if r is in the range from 225° to 315° (the palm is facing downwards) or in the range from 45° to 135° (the palm is directed upwards) the horizontal LEAP is selected as active, while the vertical is "paused" (it is still turned on, but its frame rate was reduced to the minimum in order to save computational resources). Out of these

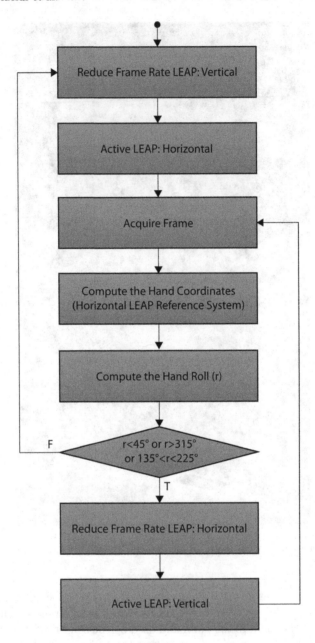

Fig. 3. The flow diagram of the tracking prototype: the hand is tracked by both sensors, the roll r with respect to the horizontal reference system is computed using the information from the active LEAP and used to determine the active sensor (image from [13]).

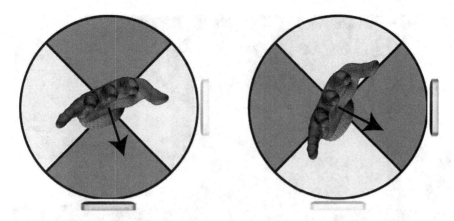

Fig. 4. The switching approach: depending on the orientation of the hand, one of the sensors is active and used to track the hand, while the other is paused, for saving computational resources and maintaining real time (image from [13]).

ranges, the vertical is active and the horizontal is paused. As explained above, the prototype behaves in a master-slave mode: the server component receives data from the routers and manipulates the information from the vertical LEAP (that is data from the Slave router) performing a roto-translation to obtain the coordinates with respect to the horizontal LEAP reference system. The server is responsible to check which LEAP is active, to send only the information received from it to the client and, if needed, to change the active status of the sensors.

For building a consistent hand model using information from both sensors, their reciprocal position has to be computed with high accuracy through a calibration procedure.

3 Calibration

The spatial positioning system contained in the numerical control mill used for drilling the boxes for the LEAPs has been also used for collecting spatial accurate measurements by using a wood stick rigidly fixed to the mill (Fig. 5). The VG support, fixed to the moving spatial system of the mill and oriented for maintaining its reference system oriented as that of the mill (the plane of horizontal LEAP was the same of the mill and this implied that also its vertical axis was in the same direction of that of the mill), was moved along the three axes with a precision of 0.01 mm while the wood stick remained fixed in its initial position. In this way it could be possible to obtain measurements inside the desired region of interest. The first measurements we collected were those regarding the positions of the vertexes of the superior surface of the two LEAPs (Fig. 6): this made it possible to evaluate their position/orientation with respect to the mill reference system (spatial calibration was obtained and orientation and positioning errors could be easily estimated and corrected without using calibration through LEAP measurements).

Fig. 5. The mill used both for producing the lodges for the LEAPs on their support and for collecting accurate spatial positioning measurements. The wood stick (a) was fixed to the static structure of the mill (b) hosting also the mill tool. The VG structure (c) was secured by a vise on the three axes moving block of the mill (d), controlled numerically. The stick was used both during the measurement of the LEAP position with respect to the mill reference system and during the assessment of spatial points with respect to the LEAPs references systems.

The resulting transformation matrices were calculated by these measurements, separately of each of the two sensors, through a Singular Value Decomposition (SVD) [13, 25–27], in homogeneous coordinates with respect to the mill coordinate system:

$$
W_{horizontal} = \begin{bmatrix} 1.0 & 8.4E-20 & 3.8E-4 & 1.2 \\ 0.0 & 1.0 & 0.0 & -181.1 \\ -3.8E-4 & 2.2E-16 & 1.0 & -4.8 \\ 0.0 & 0.0 & 0.0 & 1.0 \end{bmatrix} \tag{1}
$$

$$
W_{vertical} = \begin{bmatrix} -1.1E-16 & -1.0 & 5.9E-17 & 171.7 \\ 1.0 & -1.9E-20 & 3.2E-4 & -6.6 \\ -3.2E-4 & -5.0E-17 & 1.0 & -4.9 \\ 0.0 & 0.0 & 0.0 & 1.0 \end{bmatrix} \tag{2}
$$

However, the previous matrices taken into account just for spatial transformations of the two LEAPs and not "logical" internal differences between the

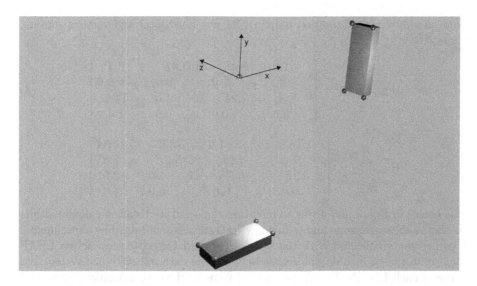

Fig. 6. Representation of the points collected by using the spatial positioning system on the corners of the LEAPS.

LEAPs and between each of their reference system and the mill reference system. For evaluating and including these effects into the previous matrices, we collected a series of spatial measurements with the first LEAP (horizontal) and then we repeated the same measurements on the same points with the second LEAP placed in place of the first. In this way, we eliminated completely eventual effects of matrix transformation on the data and difference in measurements were just due to internal differences between the two sensors.

A total of 264 measurements were collected on the surface of concentric cylinders (radii equal to: 0 cm, 2.5 cm, 5 cm, 7.5 cm and 10 cm; sampling angles: $0°$, $45°$, $90°$, $135°$, $180°$, $225°$, $270°$, $315°$) with the axis oriented along the z-axis of the LEAPs (along z, a total of 8 points were collected distributed around the centre of the VG system, distance between measurements along $z = 3$ cm, for a total length of 21 cm) in order to capture the transformation occurring from one LEAP to the other. 15 of the collected measurements, allowing to the external bases of the cylinders, were discarded because of their instabilities (we found that the z axis was really sensitive to changes in measurement: this was probably due to the difficulty of the sensor to recognize the tip of the stick along its long direction).

Also these measurements were analysed through SVD to find the resulting transformation matrices, one for each LEAP. By analysing the resulting matrices, we obtained that transformation between the coordinate system of each LEAP and the mill coordinate system was simply a translation (no distortions or scaling factors were present and just negligible fluctuations, due to measurement noise, occurred in the upper left region of the matrix). The obtained translations, were respectively included into Eqs. 1 and 2 in order to obtain the following, and final, modified transformation matrices for the horizontal and vertical sensors, respectively:

$$W_{horizontal} = \begin{bmatrix} 1.0 & 8.4E-20 & 3.8E-4 & 0.4 \\ 0.0 & 1.0 & 0.0 & -189.6 \\ -3.8E-4 & 2.2E-16 & 1.0 & 14.4 \\ 0.0 & 0.0 & 0.0 & 1.0 \end{bmatrix} \quad (3)$$

$$W_{vertical} = \begin{bmatrix} -1.1E-16 & -1.0 & 5.9E-17 & 177.4 \\ 1.0 & -1.9E-20 & 3.2E-4 & 0.8 \\ -3.2E-4 & -5.0E-17 & 1.0 & 2.9 \\ 0.0 & 0.0 & 0.0 & 1.0 \end{bmatrix} \quad (4)$$

The choice of Eqs. 3 and 4 was to maintain the accurate rotation calculated just by spatial measurements and to include the translation obtained by experimental measurements collected with the sensors (we had forced to step across LEAP measurements to include differences between the external reference system and those of each LEAP). Equations 3 and 4 were used for transforming data from both LEAPs to the mill reference system.

4   Spatial Characterization and Tracking

4.1   Spatial Characterization

The spatial characterization of the system was performed by collecting measurements on the same positions used for calibration but with the LEAPs in their own, definitive, positions. Three types of data were collected: spatial information (wood stick tip position given by the mill) and data collected by both LEAPs. Data from each LEAP were transformed and reported to the world coordinate system (the mill system) and distance between each point measured by the LEAP and the spatial measurements collected by the mill was calculated both separately for each coordinate and globally. Results were shown in Figs. 7 and 8. Pseudo-periodical trends indicate that errors increased with distance from the sensors: measurements were collected following trajectories aimed at optimizing the movements of the mill.

This produced an error reduction if the point was approaching the LEAP or an increment when it moved away. Numerically, the average error positioning (distance), standard deviation and maximum error, were reported in Table 1 separately for each LEAP. By analysing data, it can be argued that the error is bigger along the z axis and that the two LEAPs behaviour is almost the same. However, the maxim error, obtained along the z axis of each LEAP (due to the difficulty of the sensor software to indicate the tip of the stick along its long axis), was always below 7 mm. Since our scope is to identify hand tips and joints whose dimensions, for an adult person, are normally above 1 cm along each axis, our results represent a good tolerance for hand joints identification. Moreover, the use of constraints between hand joints movements and the use of temporal filtering during tracking could further improve accuracy.

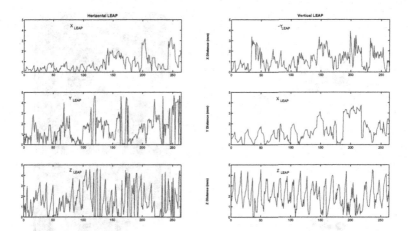

Fig. 7. The distance (expressed in mm) between points measured in different conditions along the axes for the horizontal (left panel) and vertical (right panel) sensors. Since all measurements are referred to the mill reference system, the original LEAP axes were reported inside the graphs for allowing a direct comparison between corresponding LEAP axes.

4.2 Preliminary Hand Tracking

In order to observe the real time behaviour of the VG system, a sequence of about 30 s of free hand movement was collected and the corresponding numerical hand model was represented on a computer screen. The fingers were continuously moving while the arm was rotated around the wrist back and forth. Figure 9 shows a set of hand positions and the corresponding model

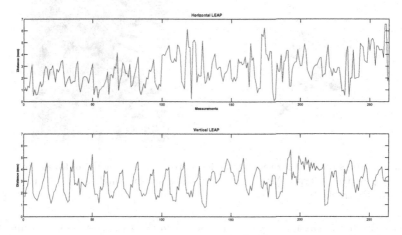

Fig. 8. The distances (expressed in mm) between points measured in different conditions for the horizontal (up panel) and the vertical (down panel) sensors.

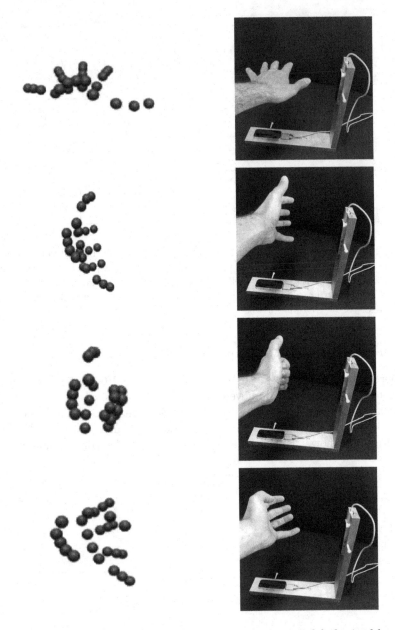

Fig. 9. Examples of hand positions with the corresponding model obtained by means of the proposed prototype. Also critical positions for the single sensor scenario have been correctly tracked.

reconstructions, obtained with the proposed approach. The accuracy and the fluidity of the tracking process were adequate (about 25 frames per second) and the change of perspective did not produce jumps or other disturbing effects.

Moreover the hand was correctly tracked also in those positions that would be critical in the single sensor scenario. Due to the fact that no hand numerical model was still associated to the virtual hand representation and that the hand was freely moving, by these measurements it was impossible to quantify the positioning error (we can reasonably argue that it could be no worse than that described above).

Table 1. Average values, standard deviations and maximum values of the distance along all the axes and in the space, between points measured by the mill and by each of the LEAPs.

		Horizontal leap	Vertical leap
X axis distance	Average	0.8	1.3
	Standard deviation	0.8	0.9
	Maximum	3.3	3.9
Y axis distance	Average	1.4	1.4
	Standard deviation	1.2	1.0
	Maximum	4.7	3.8
Z axis distance	Average	1.7	2.0
	Standard deviation	1.4	1.1
	Maximum	4.6	4.6
3D distance	Average	2.8	3.0
	Standard deviation	1.4	1.1
	Maximum	6.6	5.6

5   Conclusions

A multiple-sensor VG for real time hand tracking, based on the use of two orthogonal LEAP sensors, has been characterized and accurate positioning error measurements presented. A procedure for VG assembly and calibration has been illustrated. The average accuracy of the system is about 2–3 mm in the considered ROI, thus making the proposed system well suitable for accurate hand tracking measurements for rehabilitation purposes. This is conformed by the preliminary test of hand tracking. Future developments will regard the implementation of: (1) an efficient strategy for merging data coming from both sensors, in substitution of the actually used (mutual exclusion) strategy, for improving spatial accuracy and further reducing occlusions; (2) a numerical hand model to be associated to the virtual representation of the hand and used for forces calculation; (3) a framework for developing rehabilitation tasks associated with virtual environments and for numerical analysis of rehabilitation data and therapy outcomes. Finally, we aim at testing the VG on voluntary patients under the supervision of therapy experts.

Acknowledgments. The Authors are very grateful the Department of Physics and Chemical Sciences of the University of L'Aquila for having allowed the use of the mill and, in particular, to Mr Francesco del Grande for his invaluable help in constructing the Virtual Glove support and collecting experimental measurements.

References

1. Kopp, B., Kunkel, A., Mnickel, W., Villringer, K., Taub, E., Flor, H.: Plasticity in the motor system related to therapy-induced improvement of movement after stroke. Neuroreport **10**, 807–810 (1999). https://doi.org/10.1097/00001756-199903170-00026

2. Liepert, J., Bauder, H., Miltner, W.H.R., Taub, E., Weiller, C.: Treatment-induced cortical reorganization after stroke in humans. Stroke **31**, 1210–1216 (2000). https://doi.org/10.1161/01.STR.31.6.1210

3. Hallett, M.: Plasticity of the human motor cortex and recovery from stroke. Brain Res. Rev. **36**, 169–174 (2001). https://doi.org/10.1016/S0165-0173(01)00092-3

4. Arya, K.N., Pandian, S., Verma, R., Garg, R.K.: Movement therapy induced neural reorganization and motor recovery in stroke: a review. J. Bodywork Mov. Ther. **15**, 528–537 (2011). https://doi.org/10.1016/j.jbmt.2011.01.023

5. Burgar, C.G., Lum, P.S., Shor, P.C., Van Der Loos, H.F.M.: Development of robots for rehabilitation therapy: the Palo Alto VA/Stanford experience. J. Rehabil. Res. Dev. **37**, 663–673 (2000). http://citeseerx.ist.psu.edu/viewdoc/download?doi=10.1.1.551.4188&rep=rep1&type=pdf

6. Kahn, L.E., Lum, P.S., Rymer, W.Z., Reinkensmeyer, D.J.: Robot-assisted movement training for the stroke-impaired arm: does it matter what the robot does? J. Rehabil. Res. Dev. **43**, 619 (2006). https://doi.org/10.1682/JRRD.2005.03.0056

7. Placidi, G.: A smart virtual glove for the hand telerehabilitation. Comput. Biol. Med. **37**, 1100–1107 (2007). https://doi.org/10.1016/j.compbiomed.2006.09.011

8. Franchi, D., Maurizi, A., Placidi, G.: A numerical hand model for a virtual glove rehabilitation system. In: Proceedings of the IEEE Medical Measurement and Applications, MeMeA 2009, pp. 41–44 (2009). https://doi.org/10.1109/MEMEA.2009.5167951

9. Franchi, D., Maurizi, A., Placidi, G.: Characterization of a simmechanics model for a virtual glove rehabilitation system. In: Barneva, R.P., Brimkov, V.E., Hauptman, H.A., Natal Jorge, R.M., Tavares, J.M.R.S. (eds.) CompIMAGE 2010. LNCS, vol. 6026, pp. 141–150. Springer, Heidelberg (2010). https://doi.org/10.1007/978-3-642-12712-0_13

10. Zimmerli, L., Jacky, M., Lnenburger, L., Riener, R., Bolliger, M.: Increasing patient engagement during virtual reality-based motor rehabilitation. Arch. Phys. Med. Rehabil. **94**, 1737–1746 (2013). https://doi.org/10.1016/j.apmr.2013.01.029

11. Placidi, G., Avola, D., Iacoviello, D., Cinque, L.: Overall design and implementation of the virtual glove. Comput. Biol. Med. **43**, 1927–1940 (2013). https://doi.org/10.1016/j.compbiomed.2013.08.026

12. Llorns, R., No, E., Colomer, C., Alcaiz, M.: Effectiveness, usability, and cost-benefit of a virtual reality-based telerehabilitation program for balance recovery after stroke: a randomized controlled trial. Arch. Phys. Med. Rehabil. **96**, 418–425 (2015). https://doi.org/10.1016/j.apmr.2014.10.019

13. Placidi, G., Cinque, L., Petracca, A., Polsinelli, M., Spezialetti, M.: A virtual glove system for the hand rehabilitation based on two orthogonal leap motion controllers. In: Proceedings of the 6th International Conference on Pattern Recognition Applications and Methods, ICPRAM, vol. 1, pp. 184–192 (2017). https://doi.org/10.5220/0006197801840192

14. Lum, P.S., Godfrey, S.B., Brokaw, E.B., Holley, R.J., Nichols, D.: Robotic approaches for rehabilitation of hand function after stroke. Am. J. Phys. Med. Rehabil. **91**, S242–S254 (2012). https://doi.org/10.1097/PHM.0b013e31826bcedb

15. Maciejasz, P., Eschweiler, J., Gerlach-Hahn, K., Jansen-Troy, A., Leonhardt, S.: A survey on robotic devices for upper limb rehabilitation. J. Neuroeng. Rehabil. **11**, 10–1186 (2014). https://doi.org/10.1186/1743-0003-11-3

16. Rusk, Z., Antonya, C., Horvth, I.: Methodology for controlling contact forces in interactive grasping simulation. Int. J. Virtual Reality **10**, 1 (2011)

17. Avola, D., Spezialetti, M., Placidi, G.: Design of an efficient framework for fast prototyping of customized humancomputer interfaces and virtual environments for rehabilitation. Comput. Methods Programs Biomed. **110**, 490–502 (2013). https://doi.org/10.1016/j.cmpb.2013.01.009

18. Chaudhary, A., Raheja, J.L., Das, K., Raheja, S.: Intelligent approaches to interact with machines using hand gesture recognition in natural way: a survey. arXiv preprint arXiv:1303.2292 (2013). https://doi.org/10.5121/ijcses.2011.210

19. Placidi, G., Avola, D., Ferrari, M., Iacoviello, D., Petracca, A., Quaresima, V., Spezialetti, M.: A low-cost real time virtual system for postural stability assessment at home. Comput. Methods Programs Biomed. **117**, 322–333 (2014). https://doi.org/10.1016/j.cmpb.2014.06.020

20. Charles, D., Pedlow, K., McDonough, S., Shek, K., Charles, T.: Close range depth sensing cameras for virtual reality based hand rehabilitation. J. Assistive Technol. **8**, 138–149 (2014). https://doi.org/10.1108/JAT-02-2014-0007

21. Placidi, G., Petracca, A., Pagnani, N., Spezialetti, M., Iacoviello, D.: A virtual system for postural stability assessment based on a TOF camera and a mirror. In: Proceedings of the 3rd 2015 Workshop on ICTs for Improving Patients Rehabilitation Research Techniques, pp. 77–80 (2015). https://doi.org/10.1145/2838944.2838963

22. Leap motion inc. http://www.leapmotion.com. Accessed 2017

23. Bachmann, D., Weichert, F., Rinkenauer, G.: Evaluation of the leap motion controller as a new contact-free pointing device. Sensors **15**, 214 (2015). https://doi.org/10.3390/s150100214

24. Petracca, A., Carrieri, M., Avola, D., Basso Moro, S., Brigadoi, S., Lancia, S., Spezialetti, M., Ferrari, M., Quaresima, V., Placidi, G.: A virtual ball task driven by forearm movements for neuro-rehabilitation. In: 2015 International Conference on Virtual Rehabilitation Proceedings (ICVR), pp. 162–163 (2015). https://doi.org/10.1109/ICVR.2015.7358600

25. Sabata, B., Aggarwal, J.K.: Estimation of motion from a pair of range images: a review. CVGIP: Image Underst. **54**, 309–324 (1991). https://doi.org/10.1016/1049-9660(91)90032-K

26. Besl, P.J., McKay, N.D.: A method for registration of 3-D shapes. IEEE Trans. Pattern Anal. Mach. Intell. **14**, 239–256 (1992). https://doi.org/10.1109/34.121791

27. Eggert, D.W., Lorusso, A., Fisher, R.B.: Estimating 3-D rigid body transformations: a comparison of four major algorithms. Mach. Vis. Appl. **9**, 272–290 (1997). https://doi.org/10.1007/s001380050048

Congestion Analysis Across Locations Based on Wi-Fi Signal Sensing

Atsushi Shimada[✉], Kaito Oka, Masaki Igarashi, and Rin-ichiro Taniguchi

Kyushu University, 744, Motooka, Nishi-ku, Fukuoka, Japan
atsushi@ait.kyushu-u.ac.jp
http://limu.ait.kyushu-u.ac.jp/

Abstract. Many studies related to congestion analysis focus on estimating quantitative values such as actual number of people, mobile devices, and crowd density. In contrast, we focus on perceptual congestion rather than quantitative congestion; however, we also analyze the relationship between quantitative and perceptual congestion. We construct a system for estimating and visualizing congestion and collecting user reports about congestion. We use the number of mobile devices as quantitative congestion measurements obtained from Wi-Fi packet sensors and a user report-based congestion as a perceptual congestion measurement collected via our Web system. In our experiments, we investigate the relationship between these values. In addition, we apply Non-negative Tensor Factorization to extract latent patterns between locations and congestion. These latent features help us to understand the relationship of the characteristics among the locations.

1 Introduction

The sensing and analysis of 'people flow' is studied widely based on various sensory data such as monitoring system with stereo cameras [5], laser range finders for human tracking [9], and data-mining collected by Location-Based Services (LBS) data [6]. Although these methods provide reasonable results to understand the flow of people, there are some disadvantages. One of the important issues is how to identify people to acquire the flow. People flow analysis based on camera sensors/laser range finders has difficulty in identification of a person between different sensors since personal ID information is not available directly. In addition, these sensors are expensive and difficult to install in new environments. In another vein, people flow analysis based on LBS has poor data coverage. The quality of the analysis strongly depends on the number of users who use the application at a certain location. For instance, the Foursquare dataset[1] in New York City has 3,112 users in it, but the data consists of 0.036% of the population in New York City.

---

[1] Foursquare Dataset. https://sites.google.com/site/yangdingqi/home/foursquare-dataset/. Accessed 22 August 2016.

© Springer International Publishing AG, part of Springer Nature 2018
M. De Marsico et al. (Eds.): ICPRAM 2017, LNCS 10857, pp. 204–221, 2018.
https://doi.org/10.1007/978-3-319-93647-5_12

Recently, Probe Request sensing is gathering attention for a new approach for people flow analysis [3,11]. A Probe Request is a Wi-Fi connection request packet from a Wi-Fi device to nearby Access Points (APs). The Prove Request sensing can solve the above-mentioned disadvantages. First, it becomes easier to collect the identified flows of each person because the packet includes the device ID (called MAC address). Second, it is also easy to collect a large scale of data because Wi-Fi devices send Probe Requests periodically while the Wi-Fi is turned on. In other words, it is not necessary to install a specific application to collect data. Finally, Probe Request sensors are small and not expensive, so it is easy to install the sensing system in a new environment. (Table 1 summarizes the comparison.)

Table 1. Comparison of methods for people flow analysis.

Method	Person tracking between sensors	Data coverage	Installation
Camera	Difficult	High	Difficult
Laser-range-finder	Difficult	High	Difficult
LBS	Easy	Low	Easy
Probe request sensing	Easy	High	Easy

Congestion measurements and estimates based on people flow analysis are useful and important for various applications. For instance, it can aid in congestion avoidance and mitigation. It is also useful in ascertaining the number of visitors to retail stores for purposes of customer analysis and marketing strategies. Additionally, evacuation planning for emergency situations requires congestion information [2]. In another vein, Wi-Fi packet sensors also estimate the number of mobile devices (e.g., smartphones and laptop computers). The number of mobile devices tends to be proportional to the number of people, so we can use them to roughly estimate congestion. Wi-Fi packet sensors cover distances between a dozen to a hundred meters. A Wi-Fi radio wave has a higher transmittance than visible light, therefore we can install packet sensors in typical situations without a consideration of blind areas.

The above-mentioned techniques are aimed at estimating quantitative congestion measurements such as people count, crowd density, and the number of mobile devices. Of course, estimating the actual number of people is meaningful and important for customer analysis. Meanwhile, qualitative congestion measurements, such as a person's perception, are important rather than quantitative values when providing congestion information to people. Figure 1 shows two spots with almost the same crowd density. There are few vacant seats in the dining hall, so we would feel that the dining hall is crowded. The crowd density at the bus stop is similar to that at the dining hall, but the bus stop cannot be considered crowded. Human perception about congestion depends on

(a) Dining hall   (b) Bus stop

Fig. 1. Two spots with similar crowd density.

the people count and density and also the location fs characteristics such as area and seating capacity.

In this paper, we focus on the relationship between quantitative and perceptual congestion to provide perceptual congestion information. The number of mobile devices obtained from Wi-Fi packet sensors is used for quantitative congestion measurements. On the other hand, user report-based congestion is collected via our Web service, and it is used as a perceptual congestion measurement.

In addition, we tackle an analysis of congestion patterns. It is important issue to understand the characteristics of a particular location and relation among locations. In general, congestion data consists of high dimensional information (the number of locations, time resolutions, etc.); thus, effective dimension reduction methods are required for a better understanding of these characteristics. However, some dimension reduction methods, such as Principal Component Analysis, are not helpful for interpreting the data. One reason is that they lose the original meaning of each axis (e.g. locations, dates, time, etc.) and it is often hard to understand what each axis mean after the reduction. Specifically, we apply a Non-negative Tensor Factorization (NTF), which is a kind of dimension reduction method that does not lose the original meaning of each axis, to extract the latent congestion patterns. In our experiments, we conducted a user study, involving 304 participants, to investigate the effectiveness of our dimension reduction strategy. This paper is an extended version of the paper published in ICPRAM2016 [8]. The extended part and the main contribution with respect to the previous work is the analysis of congestion data based on NTF (described in this paragraph) and the user study (reported in the section of experimental results).

2   Related Work

2.1   People Flow Analysis Based on Wi-Fi Packet Sensing

Schauer et al. installed two Probe Request sensors at a German airport. One sensor was arranged before people pass the security check, and another after the check [11]. They analyzed the correlation between the estimated number of people by Probe Request sensing and the actual number of people that passed the security check. They demonstrated that the high correlation value (in fact, the value was 0.75 on average) through the experiment over 16 days.

Fukuzaki et al. developed their own Probe Request sensors, and applied to people flow analysis at an exhibition of Osaka Electro-Communication University over two days [4]. They analyzed the number of people and the length of time they stayed at each location. The results were utilized for the generation of origin-destination table that showed the number of people who moved from one point to the other. They concluded that the rough tendency of people flow was grasped.

In terms of counting the number of people, not only Probe Request sensing, but also the other approaches are proposed: based on the number of Wi-Fi frame [14], Channel State Information (CSI) of Wi-Fi [13], Bluetooth scan data [12].

Above-introduced studies aim at estimation of the number of people as accurately as possible based on Wi-Fi/wireless signal sensing. And, there is no work that consider the people's perception for the congestion.

2.2 Prediction and Recommendation Based on Tensor Factorization

Sahebi et al. proposed Feedback Driven Tensor Factorization, to model student learning processes and predicting student performances [7]. They created a three-dimensional tensor that indicated whether a certain student passed or failed quiz Q on a certain attempt. The tensor was, then, factorized it into another three-dimensional tensor and matrix. The three-dimensional tensor calculated from the factorization indicates students' process of acquiring knowledge (e.g., what pointers do in programming) by solving quizzes. The decomposed matrix revealed which knowledge was useful for in answering quizzes. Their approach showed higher accuracy than other approaches for the task of predicting student performance.

A mobile recommendation system was proposed to help those wishing to sightsee or dine in a large city [15]. Their system returns recommended activities at a location where a user sends a request. They proposed PARAFAC-based tensor factorization with some prior knowledge terms for this recommender system [1]. They confirmed that their approach outperformed other baseline approaches in terms of a recommendation task by comparing the accuracy of estimating the null values in the original tensor.

Oka et al. applied an NTF to people flow tensor. They added two constraints: a sparsity constraint and an initialization with prior knowledge, to the factorization process in order to help interpretation of the decomposed matrices [10]. The sparsity constraint clarifies which factor is important for some users, locations, and times. Prior knowledge (e.g., 8:00 h is breakfast time, Restaurant A is open from 08:00 h to 19:00 h, etc.) also helps our understanding of the data. They used prior knowledge by initializing the place and time latent factor matrices. By setting an initial value according to prior knowledge, they could not only examine whether extracted patterns fit to the given prior knowledge, but also discover unexpected patterns of people flow.

Above studies utilize tensor factorization techniques in order to develop a prediction/recommendation system to support human activities. In such systems, researchers do not have to consider the readability of the original and decomposed data, because the most important thing is to produce higher performance. In contrast, our study utilizes the tensor factorization to help users understand the congestion of each location and/or relationship between locations. In general, the tensor factorization is often used for dimension reduction, but its readability is not discussed in details. Therefore, we investigate the effectiveness of tensor factorization in terms of readability of the decomposed data through a user study involving 304 participants.

3   Congestion Estimate and Collecting User Reports

3.1   System Overview

An overview of our proposed system is shown in Fig. 2. When our Wi-Fi packet sensors capture probe request frames, tuples of the received time, location ID, and MAC address are stored to the database. Then, the system calculates the extent of the congestion. Our Web service plots a time series of the congestion for each location. This service has a function for receiving user reports about the congestion. The details of the system are described in following subsection.

3.2   Probe Request Capturing and Filtering

We used Wi-Fi packet sensors arranged in various locations to capture probe request frames. Wi-Fi packets can be received even when the receiver is several hundred meters away from the sender. In this study, we estimated the congestion at dining halls and bus stops on a university campus. We filtered out packets with weak signal strengths (under $-80\,\mathrm{dB}$) so that we only collected data from close devices. The received time t of the packet, place ID (sensor ID) p and MAC address m were stored to the database D^2.

$$D \leftarrow D \cup \{(t, p, m)\} \tag{1}$$

3.3   Congestion Degree Based on Probe Requests

The congestion degree $c(t, p)$ id defined for each time and place using probe request data without prior knowledge of the location as

$$c'(t, p) = |\{m \mid (t', m, p') \in D, t - 180 \le t' \le t, p' = p\}| \tag{2}$$

$$c(t, p) = \frac{c'(t, p)}{\alpha \max_t (c'(t, p))}, \tag{3}$$

---

[2] Actually, we stored hash values to the database instead of MAC addresses because of privacy issues.

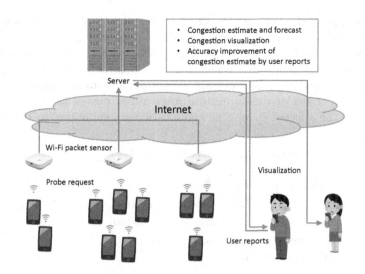

Fig. 2. Overview of our system.

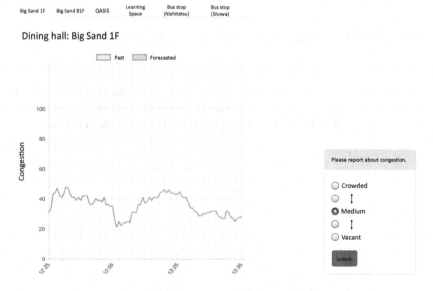

Fig. 3. Web service for visualizing congestion. (Color figure online)

Fig. 4. User report form.

where $c'(t, p)$ is the number of the unique MAC address observed during the three minutes. We obtain $c(t, p)$ by normalizing $c'(t, p)$. The value α is determined empirically (in this paper, $\alpha = 0.75$).

3.4 Visualizing Congestion and User Report

We developed a Web service to visualize the extent of the congestion. Figure 3 plots the congestion degree calculated using Eq. 3. The red line represents the congestion during the last 30 min and the green line represents the forecasted congestion. Readers can browse other locations using the upper tabs.

User reports are collected via our Web service, which are useful information as perceptual congestion measurements. Figure 4 shows the form for reporting the congestion degree on our website. There are five radio (option) buttons. Users can only select one radio button. After a user pushes the submit button on the form, the selected item, time and location are submitted to the server.

Table 2. Location of Wi-Fi packet sensors.

ID	Location	Floor	Type	Purpose
1	Dining hall A	GF	Indoor	Breakfast, lunch and dinner
2	Dining hall B	GF	Indoor	Breakfast, lunch and dinner
3	Learning space	3F	Indoor	Learning
4	Dining hall C	B1	Indoor	Lunch
17	Bus stop A	N/A	Outdoor	Returning home
18	Bus stop B	N/A	Outdoor	Returning home

4   Analysis of Congestion and User Reports

4.1   Operation of Our System

We installed six Wi-Fi packet sensors on an university campus (Figs. 5 and 6). Table 2 shows the installed spot, with an indication of whether or not the spot is located indoors. These packet sensors have been in operation since January 2016.

Fig. 5. Wi-Fi packet sensor.

We have been operating our Web service for visualizing congestion and recording user reports since July 2016. We received over three hundred user reports about congestion via our Web service during the first four weeks.

4.2   Time Series of Congestion

Figure 7 shows the congestion calculated using Eq. 3 for a typical week. We can see that the dining halls (IDs 1, 2, and 4) have a steep peak around 12:00 because of lunchtime. The congestion at the bus stops (IDs 17 and 18) tends to fluctuate intensely because busses arrive every 5 to 15 min to take passengers. Both bus stops are mainly used by people returning home, so peak congestion occurs around evening time.

4.3   Correlation Analysis of User Reports

We analyzed the correlation between user reports and the congestion recorded using Wi-Fi packet sensors. Figure 8 shows the scatter diagrams and correlation coefficients for the user reports and Wi-Fi-based congestion for each location. The correlation coefficients for locations 1, 3 and 4 are over 0.6, so we can say that the quantitative and perceptual congestion of those spots have moderate correlations. Meanwhile, the correlation coefficient for location 2 is less than 0.5 even though it is from the same category (dining hall) as locations 1 and 4. After analyzing the user reports in more detail, we found out that this low correlation due to many submissions of 'Crowded' in a short period of time. During this period, the Wi-Fi packet sensors did not estimate that the location was crowded; therefore, we believe that these submissions were malicious. We will deal with such malicious submissions in the future.

Fig. 6. Wi-Fi packet sensors on a campus.

Table 3. Time table of bus stop (location 17).

8:11 8:25 8:41   9:11   9:41 10:21 10:41 11:16 11:41 12:11 12:46 13:01
13:22 13:47 14:17 14:37 14:47 14:57 15:12 15:42 16:12 16:27 16:32 16:46
16:50 16:57 17:17 17:42 18:17 18:22 18:42 18:57 19:27 19:42 20:02 20:31
21:01 21:31 22:01

Table 4. Time table of bus stop (location 18).

6:57	6:59	7:11	7:21	7:37	7:39	7:46	7:56	8:13	8:21	8:34	8:39
8:44	8:49	8:54	8:59	9:04	9:14	9:30	9:36	9:46	9:57	10:12	10:17
10:26	10:36	10:44	10:57	11:01	11:12	11:27	11:41	11:51	11:56	12:04	12:11
12:16	12:21	12:26	12:41	12:57	13:06	13:12	13:26	13:36	13:46	13:57	14:11
14:21	14:26	14:37	14:41	14:44	14:47	14:51	14:54	14:58	15:01	15:04	15:11
15:16	15:21	15:26	15:37	15:41	15:46	15:56	16:01	16:04	16:09	16:14	16:19
16:24	16:27	16:31	16:34	16:38	16:43	16:46	16:51	16:56	17:04	17:06	17:12
17:14	17:19	17:22	17:24	17:26	17:29	17:35	17:39	17:42	17:44	17:48	17:52
17:55	18:01	18:04	18:06	18:11	18:17	18:21	18:26	18:31	18:37	18:41	18:44
18:53	18:56	18:59	19:06	19:11	19:14	19:17	19:26	19:31	19:36	19:46	19:51
19:54	19:57	20:06	20:11	20:14	20:26	20:29	20:41	20:49	21:01	21:06	21:09
21:14	21:24	21:26	21:41	21:59	22:01	22:06	22:21	22:48	22:53	22:59	23:13

The correlation coefficients of two bus stops (locations 17 and 18) are not that large (around 0.5) because of the intense fluctuations in the congestion calculated using Wi-Fi sensors. Figure 9 shows the congestion of two bus stops and Tables 3 and 4 show their timetables. Given that busses run frequently, the congestion curve fluctuates intensely; consequently, the correlation coefficients of the bus stops are not high and therefore forecasting the congestion becomes slightly difficult.

5   Spatio-Temporal Feature Analysis Across Locations

5.1   Analysis Overview

In this section, we explain the proposed framework for extracting the latent congestion patterns of people from Probe Request data. The overview of the analysis is shown in Fig. 10. In order to extract congestion patterns that indicate the number of people who stayed at each location, we compose a three-dimensional tensor that shows time, date and locations. Here, the element of the tensor indicates the number of people. In fact, the number of people is normalized to a value between 0 and 1. Table 5 shows examples of the tensor data.

We then factorize this tensor into three matrices: a location latent factor matrix, date latent factor matrix, and time latent factor matrix. Note that we can reduce the dimension of the data without losing the original meaning of each axis (location, date, and time), by applying NTF.

5.2   Non-negative Tensor Factorization (NTF) for Extracting Understandable Patterns

We use NTF to decompose the data into three matrices that indicate which factor is important for some locations, day of the week, and times. First, we show give the basis of for Tensor Factorization.

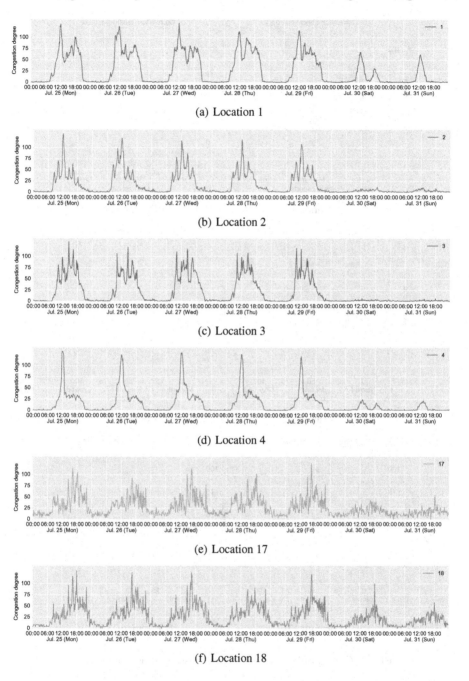

(a) Location 1

(b) Location 2

(c) Location 3

(d) Location 4

(e) Location 17

(f) Location 18

Fig. 7. Congestion for a typical week.

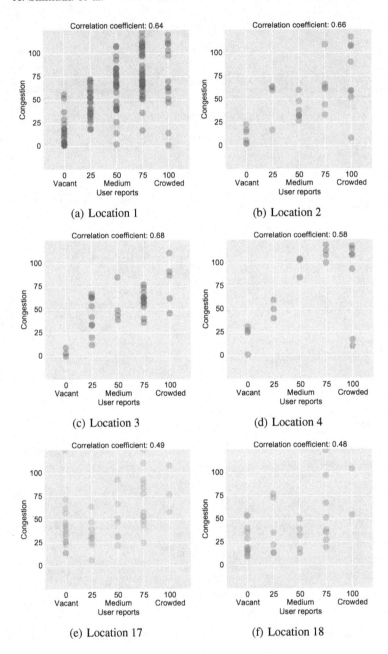

(a) Location 1   (b) Location 2

(c) Location 3   (d) Location 4

(e) Location 17   (f) Location 18

Fig. 8. Correlations between the congestion and user reports.

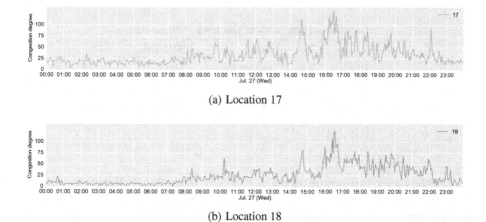

(a) Location 17

(b) Location 18

Fig. 9. Congestion of bus stops for a typical day.

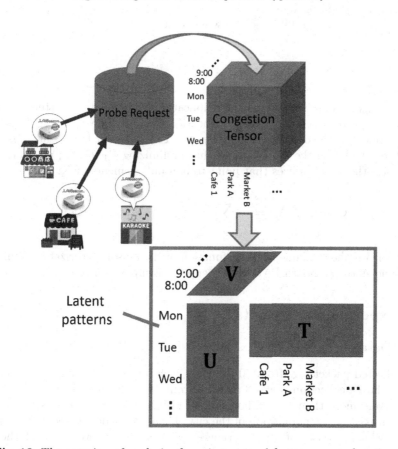

Fig. 10. The overview of analysis of spatio-temporal features across locations.

Table 5. Examples of tensor data.

Location	Time	Day of the week	Congestion
51	19	Friday	0.109188
19	20	Wednesday	0.363372
52	9	Friday	0.051829
50	2	Tuesday	0.003856
⋮	⋮	⋮	⋮

Let the target three-dimensional $L \times M \times N$ tensor be $\underline{\mathbf{X}}$. Here, we consider factorizing this $\underline{\mathbf{X}}$ into three matrices: $M \times K$ location latent matrix $\mathbf{T}$, $L \times K$ day latent matrix $\mathbf{U}$, and $N \times K$ time factor matrix $\mathbf{V}$. Note that K is a parameter that determines the number of factors. If we obtain three matrices that completely describe the original tensor $\underline{\mathbf{X}}$, then each element x_{lmn} in $\underline{\mathbf{X}}$ and each latent pattern vector $\mathbf{u}_l = [u_{l1}, \ldots, u_{lK}]^T$, $\mathbf{t}_m = [t_{m1}, \ldots, t_{mK}]^T$, and $\mathbf{v}_n = [v_{n1}, \ldots, v_{nK}]^T$ fulfills the following equation.

$$x_{nml} = \sum_{k=1}^{K} t_{lk} u_{mk} v_{nk}. \tag{4}$$

That is, x_{nml} is expressed by as a multiplication of three latent pattern vectors: the latent pattern vectors of user l, location m, and time n. Using Eq. 4, we formulate define the cost function as $C_{TF}(\mathbf{U}, \mathbf{T}, \mathbf{V})$. Tensor factorization is then equal to calculating the $\mathbf{U}$, $\mathbf{T}$, and $\mathbf{V}$ that minimizing $C_{TF}(\mathbf{U}, \mathbf{T}, \mathbf{V})$. Here, $\mathcal{D}_{\underline{\mathbf{X}}}$ is denotes the set of indices that point to non-null elements in $\underline{\mathbf{X}}$.

$$C_{TF}(\mathbf{U}, \mathbf{T}, \mathbf{V}) = \sum_{(l,m,n) \in \mathcal{D}_{\underline{\mathbf{X}}}} \left(x_{lmn} - \sum_{k=1}^{K} t_{lk} u_{mk} v_{nk} \right)^2 \tag{5}$$

Equation 5 is the fundamental cost function of the tensor factorization, which is the same as the standard PARAFAC tensor decomposition [1].

6   Experimental Results

6.1   Dataset

We collected packet logs from 40 locations in Kyoto city, Japan. The period of for the data a collection is was 3 months, from Jul. 1, 2016st to Sep. 30, 2017 .6. There were more than a 1 million unique MAC addresses, which is equivalent to about 68% of the population of the city. Then, we made created a congestion tensor which consists of 40 (locations) × 7 (day of the week) × 24 (hours) elements.

6.2   Latent Patterns

Figure 11 shows visualization results of three decomposed (latent) matrices with 5 features. In each matrix, the brightness of each cell indicates the strength of the response for a feature; the darker the color, the higher the response. For example, location 1 has a higher response to feature 2 in the location latent matrix. In the day latent matrix, feature 2 has strong relations with weekdays. In fact, there is an office of a company at location 1 and the feature indicates that the employees work on weekdays.

A hotel at location 19 has a strong relation with the feature 3 and feature 4. These features correspond to congestion around 9 AM (check-out time) and 18 PM (check-in time) respectively. Feature 0 has a strong relation with the midnight period. This feature grasps the characteristics of a bar open until late at night.

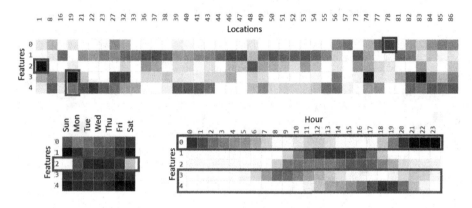

Fig. 11. Three latent matrices with 5 features.

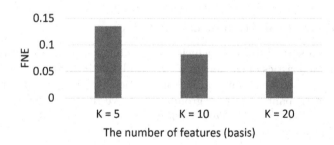

Fig. 12. Frobenius norm error (FNE) of each setting.

A recomposition error, calculated by the following formula, is often used to discuss the accuracy of the decomposition.

$$\mathrm{FNE}(\underline{\mathbf{X}}, \mathbf{T}, \mathbf{U}, \mathbf{V}) = \frac{\sum_{(l,m,n) \in \mathcal{D}_{\underline{\mathbf{X}}}} \left(x_{lmn} - \sum_{k=1}^{K} t_{lk} u_{mk} v_{nk} \right)^2}{\sum_{(l,m,n) \in \mathcal{D}_{\underline{\mathbf{X}}}} x_{lmn}^2}. \tag{6}$$

In general, as the number of features K increases, the decomposition error of the tensor becomes smaller as shown in Fig. 12. However, visibility and readability of the decomposed matrices and relations among them become lower. Figures 13, 14 and 15 show the location latent matrices when the number of features were set to 5, 10 and 20 respectively. It is not easy to understand the correlation of features across locations. Therefore, we had to determine the appropriate number of features to ease the analysis. In the next section, we report a user study.

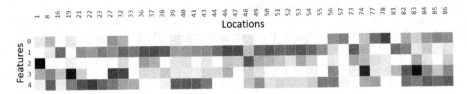

Fig. 13. Location latent matrix with 5 features.

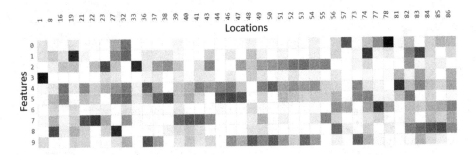

Fig. 14. Location latent matrix with 10 features.

6.3 Subjective Evaluation of Latent Patterns

We conducted a survey to assess which the latent matrix that would be ease to grasp the characteristics of the locations. In total, 304 individuals, including old and young people, old and young alike, join participated in our experiments. We divided the participants into four groups, and gave different materials to each group as denoted shown in Table 6. We asked the participants to read the given materials and fill out their findings on the answer sheet.

Table 6. Description of four groups in our survey. Description of four groups in our survey.

Group 1	The original data on a spreadsheet
Group 2	Three latent matrices with 5 features
Group 3	Three latent matrices with 10 features
Group 4	Three latent matrices with 20 features

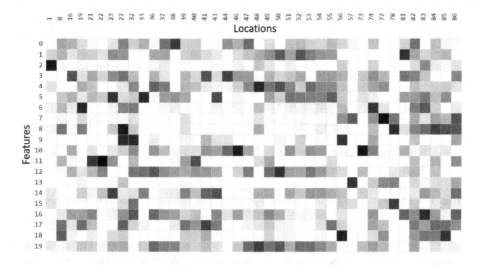

Fig. 15. Location latent matrix with 20 features.

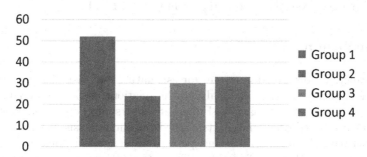

Fig. 16. The number of participants who filled out characteristics of individual location.

First, we counted the number of participants who filled out the characteristics of individual locations. The result is shown in Fig. 16. The participants in Group 1 seemed to find the characteristics of each location. This group directly read the original tensor data organized on a spreadsheet. We assume that the original data was easy to use to find location-based characteristics by extracting a specific location on the spreadsheet. In contrast, in the cases of the decomposition data (Group 2, 3 and 4), the characteristics have to be considered through the extracted features.

Second, Fig. 17 shows the number of people who filled out the relationship among locations. In contrast to Fig. 16, the number of participants in Group 1 was comparably smaller. In fact, it is not easy to discover a relation and/or correlation between even two locations from the original tensor data since there are 40 locations with 7 (day of the week) *times* 24 (hour) congestion information. On the other hand, factorization results tell us the characteristics of locations as

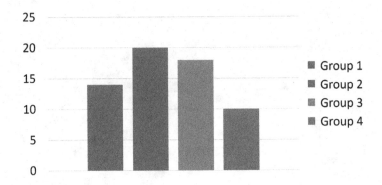

Fig. 17. The number of participants who filled out characteristics across locations.

smaller number of latent features. That is why the participants in Group 2 and Group 3 could fill out more answers in our survey. With regard to Group 4, the result indicates that it was difficult for participants to find the characteristics, as can be judged from the chaotic representation in Fig. 15.

7   Conclusion

In this paper, we described a system for estimating and visualizing congestion using Wi-Fi packet sensors. We analyzed the relationship between quantitative congestion measurements using Wi-Fi packet sensors and perceptual congestion measurements based on user reports. Based on our analysis, we found correlations between the quantitative and perceptual congestion measurements for each location. We then applied the NTF to the congestion tensor to analyze the characteristics of individual location and across locations. Through our user study, we found out that the factorization results helped people to understand the characteristics of locations.

In our future work, we plan to install Wi-Fi packet sensors at more locations (e.g., lecture rooms, laboratories, and conference rooms) and then analyze the congestion in more detail. Based on the relationship between quantitative and perceptual congestion, we will improve the accuracy of congestion estimates and provide congestion information via our system. In addition, we will collaborate with POS data for furthermore analysis of locations.

References

1. Bro, R.: PARAFAC. Tutorial and applications. Chemometr. Intell. Lab. Syst. **38**, 149–171 (1997)
2. Choi, J., Hwang, H., Hong, W.: Predicting the probability of evacuation congestion occurrence relating to elapsed time and vertical section in a high-rise building. In: Peacock, R., Kuligowski, E., Averill, J. (eds.) Pedestrian and Evacuation Dynamics, pp. 37–46. Springer, Boston (2011). https://doi.org/10.1007/978-1-4419-9725-8_4

3. Fukuzaki, Y., Nishio, N., Mochizuki, M., Murao, K.: A pedestrian flow analysis system using Wi-Fi packet sensors to a real environment. In: ACM International Joint Conference on Pervasive and Ubiquitous Computing (2014)
4. Fukuzaki, Y., Mochizuki, M., Murao, K., Nishio, N.: A pedestrian flow analysis system using Wi-Fi packet sensors to a real environment. In: Proceedings of the 2014 ACM International Joint Conference on Pervasive and Ubiquitous Computing: Adjunct Publication, pp. 721–730. UbiComp 2014. Adjunct, ACM, New York (2014). https://doi.org/10.1145/2638728.2641312
5. Heikkilä, J., Silvén, O.: A real-time system for monitoring of cyclists and pedestrians. Image Vis. Comput. **22**, 563–570 (2004)
6. Hsieh, H.P., Li, C.T., Lin, S.D.: Exploiting large-scale check-in data to recommend time-sensitive routes. In: ACM SIGKDD International Workshop on Urban Computing, UrbComp 2012 (2012)
7. Sahebi, S., Lin, Y.-R., Brusilovsky, P.: Tensor factorization for student modeling and performance prediction in unstructured domain. In: 9th International Conference on Educational Data Mining, EDM 2012 (2016)
8. Igarashi, M., Shimada, A., Oka, K., Taniguchi, R.: Analysis of Wi-Fi-based and perceptual congestion. In: 6th International Conference on Pattern Recognition Applications and Methods, ICPRAM 2017, February 2017
9. Jung, E.J., Lee, J.H., Yi, B.J., Park, J.Y., Yuta, S., Noh, S.T.: Development of a laser-range-finder-based human tracking and control algorithm for a marathoner service robot. IEEE/ASME Trans. Mech. **19**, 1963–1976 (2014)
10. Oka, K., Igarashi, M., Shimada, A., Taniguchi, R.: Extracting latent behavior patterns of people from probe request data: a non-negative tensor factorization approach. In: 6th International Conference on Pattern Recognition Applications and Methods, ICPRAM 2017, February 2017
11. Schauer, L., Werner, M., Marcus, P.: Estimating crowd densities and pedestrian flows using Wi-Fi and Bluetooth. In: 11th International Conference on Mobile and Ubiquitous Systems: Computing, Networking and Services, MOBIQUITOUS 2014 (2014)
12. Weppner, J., Lukowicz, P.: Bluetooth based collaborative crowd density estimation with mobile phones. In: 2013 IEEE International Conference on Pervasive Computing and Communications, PerCom, pp. 193–200, March 2013
13. Xi, W., Zhao, J., Li, X.Y., Zhao, K., Tang, S., Liu, X., Jiang, Z.: Electronic frog eye: counting crowd using WiFi. In: IEEE Conference on Computer Communications, INFOCOM 2014, pp. 361–369. IEEE, April 2014
14. Yaik, O.B., Wai, K.Z., Tan, I.K., Sheng, O.B.: Measuring the accuracy of crowd counting using Wi-Fi probe-request-frame counting technique. J. Telecommun. Electron. Comput. Eng. (JTEC) **8**(2), 79–81 (2016)
15. Zheng, V.W., Zheng, Y., Xie, X., Yang, Q.: Towards mobile intelligence: learning from GPS history data for collaborative recommendation. Artif. Intell. **184**, 17–37 (2012)

Text Line Segmentation in Handwritten Documents Based on Connected Components Trajectory Generation

Insaf Setitra[✉], Abdelkrim Meziane, Zineb Hadjadj, and Nawfel Bengherbia

Research Center on Scientific and Technical Information Cerist, Algiers, Algeria
{isetitra,ameziane,zhadjadj,nbengherbia}@cerist.dz

Abstract. Text line segmentation in handwritten documents is an important step in many high level processing such as handwritten document enhancement and text recognition. In this paper we describe a novel approach of text line segmentation based on tracking. In this sense, we consider each connected component in the image as a moving object in its respective line and find its best match given its history motion, i.e. the closest connected component that lie in its trajectory. Direction of motion gives direction of handwritten text and is the output of our tracking algorithm. We apply our approach to images of ICDAR 2013 handwritten segmentation contest and report an overall detection rate of 86.51%.

Keywords: Text line segmentation · Handwritten · Tracking
Connected component analysis · Trajectory generation

1 Introduction

Text line segmentation is the process of finding lines in a document image. This task is more challenging in manuscript images since writing style is different for each writer which affects skew and adjacency of text lines in the document. Among works that try to detect automatically text line in handwritten document, Li et al. [1] propose an approach for handwritten text lines segmentation using level sets. Goto and Aso [2] proposed a local linearity based method to detect text lines in English and Chinese documents. In the method proposed by Hones and Litcher [3], text lines are generated by expanding the line anchors of the document image. The previously cited methods cannot handle variable sized text, which is the main drawback.

Roy et al. [4] proposed text line extraction using foreground and background information. Louloudis et al. [5] used a block-Based Hough Transform for text line extraction. In the method proposed by Loo and Tan [6] the irregular pyramids are used for text line segmentation. Recently, Bukhari et al. [7] proposed a line segmentation approach for camera-based warped documents using active contour models. Gatos et al. [8] proposed an algorithm based on text line and

© Springer International Publishing AG, part of Springer Nature 2018
M. De Marsico et al. (Eds.): ICPRAM 2017, LNCS 10857, pp. 222–234, 2018.
https://doi.org/10.1007/978-3-319-93647-5_13

word detection for warped documents. Bai et al. [9] used a traditional perceptual grouping-based algorithm for extracting curved lines. Pal and Roy [10] proposed a head-line based technique for multi-oriented (printed in several orientations) and curved text lines extraction from Indian documents. In other work, Pal et al. [11] developed a system for English multi-oriented text line extraction estimating the equation of the text line from the character information.

This paper describes a new approach inspired of tracking works to detect lines in handwritten document images.

The approach is an extension of our previous work [12]. Each cluster of connected pixels (which can be a word or a part of a word) is considered as an object moving from left to right in the manuscript image. Its trajectory is calculated based on a regular motion. We define a regular motion of a cluster as seeking its best match among the other clusters with minimal angle deviation with respect to its current trajectory. We benefit from tracking rationale where we predict the next position of clusters according to their trajectories determined by their previous positions. We avoid tracking issues since we do not count on any feature of the cluster. That is because our moving object (the cluster representing a word or parts of a word in the handwritten document image) need not to have a similar shape in all positions of its trajectory.

Unlike our previous work in [12] where clusters were considered neighbours if they resided inside a circle centred at the current cluster, we search, in this paper, for neighbours inside a rectangular window positioned to the right of the current column coordinate. The second difference is that we explicitly generate trajectories whereas we previously only performed pair matching without creating complete trajectories. This enabled us to extend our experiment to a quantitative analysis using the metrics defined in [13] and computed by their software package. Throughout this paper, we refer to clusters as connected components. We explain more deeply our approach in Sect. 2. We present our quantitative analysis in Sect. 3 and conclude in Sect. 4 with some perspectives.

2 Our Approach

Our algorithm takes as input a binary manuscript image and produces as output a set of paths tracking the lines of written text. We first extract all connected components of the image and represent them with the coordinates (raw and column indices) of their centers of gravity (yellow dots in Fig. 1.). Components smaller than a certain threshold are deleted in order not to mislead our algorithm. Examples of such regions include dots (dark pink dots in Fig. 1.). We then generate a set of tracks (lines in the manuscript) passing by the components' centers as to minimize the distance between each two consecutive points of each track and the angle between each sequential track segments.

Formally, let $X = \{x^1, x^2, ...x^n\}$ be a list of connected component centers s.t. $x^i = \begin{pmatrix} x_1^i \\ x_2^i \end{pmatrix}$ representing the coordinates of x^i. X is sorted in the increasing

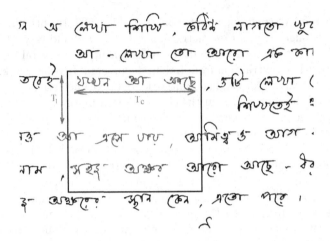

Fig. 1. Example of a center tracking. Red region: connected component to be tracked and its center highlighted in yellow. Gray rectangle: search window of the tracked connected component. The search window includes small connected components to be removed (pink regions) and connected component to be considered for comparisons and matching (green regions). Search window starts from column coordinates of the center to be tracked, and its line minus the line threshold T_l. It ends at the row and column coordinates of the center plus line and column thresholds T_l and T_c respectively. Black regions are connected components far from the connected component to be tracked and are not considered for comparisons. Yellow dots are centers of connected components. (Color figure online)

order of raw values as follows:

$$\forall i, j, i < j; x_1^i \leq x_j^i \tag{1}$$

Our algorithm generates for each connected component center a trajectory starting from it. This process is explained in the following section.

2.1   Preprocessing

Our algorithm takes as input a binary manuscript image and produces as output a set of paths tracking words or parts of a word in each line of the written text. Binary manuscript images can be issued from binarization processes. In our case, we assume binarization has already been performed and accept as input binary images where each connected component in the binary image is either a letter, a word or a part of a word depending on the writing script. We refer to a connected component as set of pixels having same value and connected with a 8−neighbouring connection.
More formally:
Let B be a binary image of a manuscript where:

$$\begin{cases} B(u,v) = 1 & if \ B(u,v) \in foreground \\ B(u,v) = 0 & if \ B(u,v) \in background \end{cases} \tag{2}$$

where u and v are row and column coordinates of an arbitrary pixel in B. We refer to foreground and background as text non text regions respectively in B. The pixel (u, v) is connected to the pixel (u', v') if there is a path:
$P = \{(u, v); (u_1, v_1); ...; (u_n, v_n); (u', v')\}$ With:

$$\begin{cases} B(u_i, v_i) = 1 & for\ i = 1; ...; n, \\ (u_{i-1}, v_{i-1}) \in N_8(u_i, v_i) & for\ i = 1; ...; n. \end{cases} \tag{3}$$

where

$$N_8((u_i, v_i)) = \begin{cases} (u_i - 1; v_i), (u_i + 1; v_i), (u_i; v_i - 1), (u_i; v_i + 1), \\ (u_i - 1; v_i - 1), (u_i - 1; v_i + 1), (u_i + 1; v_i - 1), \\ (u_i + 1; v_i + 1) \end{cases} \tag{4}$$

is the $8-$ neighbourhood of pixel (u_i, v_i).

A connected component CC is a set of pixels such that every pair of pixels in the set are connected. Once connected components extracted, we remove irrelevant connected components. Dots, being a connected component and which are parts of words, are above letters in many scripts. In the case of manuscript images, dots are even further than only at above proximity from the letter. This makes a potential tracking of a dot erroneous and can alter the whole tracking process. To overcome this issue, we remove all connected components having approximately the size of a dot, which makes our line tracking ignores dots and similar connected components. To do so, we define manually a minimum area threshold T_a and remove all connected components below this threshold: Let $CC = CC^1, CC^2, ..., CC^k, ..., CC^m$ be the set of all connected components of B. For each connected component CC^k of image B where $CC^k = \{((u_1, v_1), ..., (u_n, v_n)\}$

$$\begin{cases} if & \dfrac{\sum_{i=1}^{n}(B(u_i, v_i))}{n} < T_a : & remove\ CC^k\ from\ CC \end{cases} \tag{5}$$

After small regions removal, we compute for each connected component CC^k, its center of mass:

$$\begin{cases} x_1^k = \dfrac{\sum_{i=1}^{n} u_i}{n} \\ x_2^k = \dfrac{\sum_{i=1}^{n} v_i}{n} \end{cases} \tag{6}$$

where x_1^k, x_2^k are row and column coordinates of connected component CC^k. $C = \{C^1, ..., C^n\}$ are all the n centers of image B.

Figure 1 shows extracted connected components of image B. Centers computed in of Eq. 6 are represented by yellow dots. Removed connected components of Eq. 5 are represented by dark pink dots.

2.2   Connected Components Tracking

In order to find text lines in a manuscript, we track connected component center of each cluster CC^k found in the previous step. Let $x^k = \begin{pmatrix} x_1^k \\ x_2^k \end{pmatrix}$ be the center of cluster C^k for which we want to find the trajectory t^k, where x_1^k, x_2^k are respectively row and column coordinates of center x^k. Tracking is performed as follows

Step 1: Initialization

- The trajectory t^k is a list and is first initialized to the element x^k, i.e. $t^k = \{x^k\}$.

- We define $slopevector(x^k)$ to be the vector representing the slope of the current trajectory t^k at the point x^k. For the first point of the current trajectory x^k, we set $slopevector(x^k)$ to the horizontal unit vector $u_1 = \begin{pmatrix} 0 \\ 1 \end{pmatrix}$. That is, we assume that the trajectory starts horizontal.

Step 2: Search Window Construction

- A window starting from x^k is constructed so that to minimize the search window of the next track of x^k. Let W^k be the search window of x^k. W^k is computed as follows:

$$
\begin{cases}
W^k = \{x^i\} \ s.t. \qquad \forall x^i \in C : \\
\qquad x_1^i \leq x_1^k + T_l \\
\qquad\qquad and \\
\qquad x_1^i \geq x_1^k - T_l \\
\qquad\qquad and \\
\qquad x_2^i \leq x_2^k + T_c \\
\qquad\qquad and \\
\qquad\qquad x_2^i \geq x_2^k
\end{cases}
\tag{7}
$$

if $W^k = \emptyset$ then, tracking of x^k is stopped and go to step 5.

Step 3: Next Track Selection

- We set the next point y of the trajectory t^k to the component $x^j \in C$ which minimizes the angle between the vectors $slopevector(x^k)$ and $\overrightarrow{x^k x^j}$ ($\overrightarrow{x^k x^j}$ is the vector $x^j - x^k$.).

$$
y = \arg\min_{x^j \in neighborhood(x)} \theta(slopevector(x^k), \overrightarrow{x^k x^j})
\tag{8}
$$

The motivation behind using this minimization is to generate regular-looking trajectories with small angle between each sequential trajectory segments. The angle calculation is given by the following formula derived from the expression of the dot product between the two vectors:

$$anglebetween(\overrightarrow{u}, \overrightarrow{v}) = arcos\left(\frac{\overrightarrow{u}.\overrightarrow{v}}{\|\overrightarrow{u}\|\|\overrightarrow{v}\|}\right) \tag{9}$$

- new trajectory is then equal to previous trajectory updated with the new center: $t^k = \{t^k \cup x^j\}$.

Step 4: Tracking Update

- The slope vector is updated so that the writing style is respected and does not follow only a horizontal writing. The variable $slopevector(x)$ is updated as follows:

$$slopevector(x) \leftarrow y - x \tag{10}$$

- remove x^k from C.
- x^k is also updated to x^j so that we continue the previous tracking starting from best match.

Step 5: Loop into Remaining Tracks

- if $C = \emptyset$, go to step 6.
- else if $W^k = \emptyset$ then initialize x^k to a center in C, create a new trajectory $t^k = x^k$ and go to step 2.
- else go to step 2 with already updated variables.

Step 6: Terminate Tracking

- The tracking is terminated when all centers in C have been tracking. This results in a number of trajectories equal to connected components centers in the image B. Note that, this method generates redundant trajectories. For example, even when a center x^k is tracked to a center x^j, then, our approach generates a two trajectories while both are merged to same one. While this method is not optimized, we propose in the following how to merge trajectories.

2.3   Merging Nearby Trajectories

As discussed earlier, for each center in C we generate a new trajectory regardless of common tracks with previous trajectories. This leads to a huge number of trajectories (this number is equal to number of connected components in image B). In order to reduce this number of trajectories, we merge all trajectories which have common elements.

Let $T = \{t^1, t^2, ..., t^n\}$ be the set of n trajectories generated after tracking. For each pair $(t^i, t^j) \in T$ if $t^i \cap t^j \neq \emptyset$ then update T to $T = (T \setminus \{t^i, t^j\}) \cup T'$ and $T' = t^i \cup t^j$. After merging nearby trajectories, we affect a same value to each trajectory in T, for example, for trajectory $t^1 = \{x^1, x^2, x^3\}$, all pixels centered at $\{x^1, x^2, x^3\}$ are given the value 1. This value is also propagated to all connected components from which centers were extracted. For the same previous example, all pixels of the three connected components centered at $\{x^1, x^2, x^3\}$ are given the value 1.

2.4   Small Regions Label Propagation

In Sect. 2.2 we discussed how we removed small connected components so that our tracking is not affected. Once tracking performed, small connected components must be assigned to a text line, i.e. a previously computed trajectory. To do so, for each pixel in ignored connected components, we look for the nearest neighbour which had a label i.e. a trajectory number. This label is then propagated to this pixel. Note that by doing this, pixels of the same connected component can belong to different manuscript lines based on proximity.

3   Experimental Results

3.1   Dataset Used

We tested our approach on images of ICDAR 2013 Handwriting Segmentation Contest, [13]. The dataset consists of 150 document images written in English and Greek as well as 50 images written in Bangla along with the associated ground truth for training and 50 images written in English, 50 images written in Greek and 50 images written in Bangla for test.

3.2   Metrics Used

In order to assess numerically accuracy of the approach, we compute the detection accuracy (DR) metric proposed in [13]. DR of image B computed as follows:

$$DR = \frac{o2o}{N} \tag{11}$$

where $o2o$ and N are respectively the one to one match and number of ground truth manuscript lines. In order to compute $o2o$ we first need to compute the Match Score (MS) of the two regions represented by their labels i and j as follows:

$$MS(i, j) = \frac{G_j \cup R_i \cup B}{(G_j \cup R_i) \cap B} \tag{12}$$

G_j is is a set of pixel coordinates in B where pixels share same label i corresponding to a manuscript line in the ground truth. R_i is the set of pixel coordinates in B where pixels share same label j corresponding to a trajectory in the resulting trajectories sets. MS is computed for all (j, j) pairs because, manuscript line with label 1 in the ground truth for example, can be different of manuscript line with label 1 is the resulting trajectories. In order to compute $o2o$, we first initialize $o2o = 0$, then for each tuplet resulting region and ground truth region (i, j), we check if $MS(i, j) \geq T_{acc}$, in this case, the two regions are accepted as a one to one match, $o2o = o2o + 1$. In the best case, $o2o$ will be equal to number of ground truth regions, which results in a DR of 1 (or 100% as percentage).

In our implementation, we first use the framework proposed in which provides computation of several metrics. We leave the acceptance threshold T_{acc} by default i.e. $T_{acc} = 0.9$, and compute DR. We then, reimplement the metrics and check exactitude with results of the framework. We do so, because we would like to analyse impact of our approach parameters, namely T_l (line threshold) and T_c (column threshold) on the detection rate DR.

3.3 Parameters Tuning

Our approach requires only three parameters: Threshold Area (T_a) that controls small regions removal, line threshold T_l and column threshold T_c that define search window size. Parameters are described in Sect. 2.2. It was not tedious to choose T_a as we chose it according to image size and fix it to $T_a = 150$. Images of the dataset being high resolution, connected components of all images were relatively high, we chose T_a as a reasonable threshold. The two remaining thresholds had high impact on accuracies. In order to inspect impact of these two parameters and choose their best values, we propose to loop the tracking process over different values of T_l and T_c.

Table 1. Detection rate (DR) by varying T_l and T_c.

T_i	20	20	20	20	40	40	40	40	60	60
T_c	100	250	400	550	100	250	400	550	100	250
DR	0.96	21.60	25.65	23.24	0.98	25.59	27.97	27.34	0.98	25.25
T_i	60	60	80	80	80	80	100	100	100	100
T_c	400	550	100	250	400	550	100	250	400	550
DR	26.43	26.76	0.98	22.99	24.11	24.60	0.85	10.07	12.37	12.74

Table 2. Detection rate (DR) by varying T_l and T_c independently.

T_l	DR	T_l	DR
20	17.86	100	0.95
40	20.47	250	21.10
60	19.86	400	23.31
80	18.17	550	22.94
100	9.01		

Table 3. Detection rate (DR) for groups of images, third column is the number of images that have their best DR with same parameters.

T_l	T_c	N^o images	DR
20	200	3	83.33
20	250	4	100.00
20	400	5	98.75
20	500	2	100.00
20	550	3	98.33
40	250	21	86.54
40	400	17	92.80
40	550	5	79.08
60	250	2	84.21
60	350	18	71.91
60	400	3	85.96
60	500	4	94.12
60	550	1	83.33
100	350	2	69.16

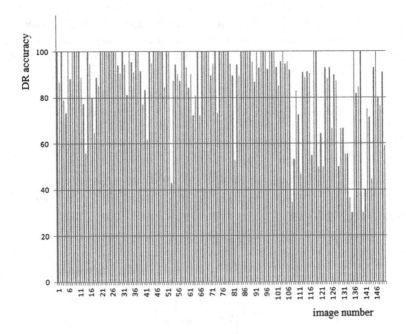

Fig. 2. DR score of our approach on images of ICDAR 2013 Handwriting Segmentation Contest. Overall detection rate is 86.51%.

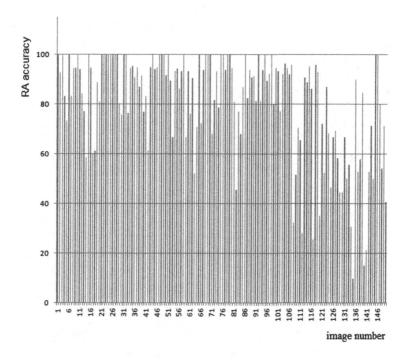

Fig. 3. RA score of our approach on images of ICDAR 2013 Handwriting Segmentation Contest. Overall detection rate is 81.00%.

We define for T_l an interval $= [20, 40, 60, 80, 100]$ and for T_c the interval $= [100, 250, 400, 550]$. This results in $4 \times 5 = 20$ combinations. For each tuplet, we first compute the mean detection rate (DR) over a subset of 90 images from the dataset. In order to have a better understanding of contribution of each threshold a part, we further compute the mean detection rate for each threshold separately.

From Table 1, it is hard to find a universal tuplet that give a good DR for all images based on T_l and T_c combination. This can be observed by the DR which is has a maximum mean for all images of no more than 30%. In Table 2, we compute again the mean DR for all images by fixing only one threshold. We observe that results are not better than previous ones. We assume here that fixing same thresholds for all images is responsible of low detection rates, and assume also that if for each image or group of images, we use different thresholds, then, DR might be improved. In order to verify this assumption, we choose for each image, thresholds that gave the best DR and keep this latter. We observe in this case a detection rate of $DR = 90.49\%$. We show in Table 3 chosen thresholds for groups of images. In the third column of the table we show number of images that share high detection rates for the same thresholds. Doing this, we can choose thresholds for groups of images. Although this method allowed us to have good detection rates, we believe that a further analysis would

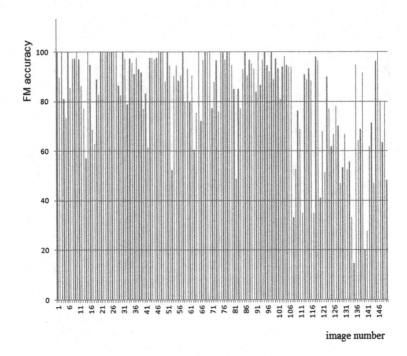

image number

Fig. 4. FM score of our approach on images of ICDAR 2013 Handwriting Segmentation Contest. Overall detection rate is 83.27%.

be good, such as analysing other relations between groups of images instead of comparing with ground truth images (distance between connected components, mean and standard deviation of connected components areas, etc.). We leave this perspective to future works.

3.4 Final Results

We implemented and tested our approach in matlab 2015. While we used only detection rate (DR) as a condition to choose best line and column thresholds for image grouping, we used directly the software provided by [13] for other metrics computation (Recognition accuracy (RA) and Final Matching Score (FM)). These metrics can be found in the referenced paper. We show results of our algorithm applied to each image of the dataset in Figs. 2, 3 and 4. Overall accuracy for the 150 images is: 86.51%, 81.00% and 83.27% for DR, RA and FM respectively. From experiments, we notice high accuracies for images where connected components are sufficiently far from each other. While our algorithm gave more regions than ground truth regions, RA Score is decreased compared to DR score. However, for all the metrics, a decrease is observed for latest images (between 131 and 141). These images are images where connected components are merged together, and where writing direction is changed drastically. Although our algorithm succeeds for getting trajectories when motion changes smoothly, it breaks when there is an abrupt change in the direction of motion. We aim in future work to add a processing where trajectories are corrected which can handle the first issue, and would like to combine our approach to a global approach in order to correct our local approach based on tracking that causes the second issue. Also, when two trajectories have at least one common component, they are merged as one line. Due to this, connected components of different lines are merged into the same trajectory. This problem can be solved when merges and splits are detected in trajectories and corrected at their specific tracks [14].

4   Conclusion

In this work, we presented a new approach for handwritten text line segmentation inspired of various tracking approaches. The aim of the approach is to track each connected components in the handwritten document from its beginning (which can be at any point in the image) until its end (which is at the right end of the image). Our approach is robust to skew and text orientation since we keep history of connected components' positions along their trajectories. However, our approach fails when common connected components of different lines cross. Trajectory analysis can solve this issue but out of the scope of this paper, is left as perspective.

References

1. Li, Y., Zheng, Y., Doermann, D., Jaeger, S., Li, Y.: Script-independent text line segmentation in freestyle handwritten documents. IEEE Trans. Pattern Anal. Mach. Intell. **30**, 1313–1329 (2008)
2. Goto, H., Aso, H.: Extracting curved text lines using local linearity of the text line. Int. J. Doc. Anal. Recogn. **2**, 111–119 (1999)
3. Hönes, F., Lichter, J.: Layout extraction of mixed mode documents. Mach. Vis. Appl. **7**, 237–246 (1994)
4. Roy, P.P., Pal, U., Lladós, J.: Text line extraction in graphical documents using background and foreground information. Int. J. Doc. Anal. Recogn. (IJDAR) **15**, 227–241 (2012)
5. Louloudis, G., Gatos, B., Halatsis, C.: Text line detection in unconstrained handwritten documents using a block-based hough transform approach. In: Proceedings of the Ninth International Conference on Document Analysis and Recognition - Volume 02, ICDAR 2007, pp. 599–603. IEEE Computer Society, Washington (2007)
6. Loo, P.K., Tan, C.L.: Word and sentence extraction using irregular pyramid. In: Lopresti, D., Hu, J., Kashi, R. (eds.) DAS 2002. LNCS, vol. 2423, pp. 307–318. Springer, Heidelberg (2002). https://doi.org/10.1007/3-540-45869-7_36
7. Bukhari, S.S., Shafait, F., Breuel, T.M.: Segmentation of curled textlines using active contours. In: The Eighth IAPR International Workshop on Document Analysis Systems 2008, DAS 2008, pp. 270–277 (2008)
8. Gatos, B., Pratikakis, I., Ntirogiannis, K.: Segmentation based recovery of arbitrarily warped document images. In: Ninth International Conference on Document Analysis and Recognition (ICDAR 2007), vol. 2, pp. 989–993 (2007)

9. Bai, N.N., Nam, K., Song, Y.: Extracting curved text lines using the chain composition and the expanded grouping method (2008)
10. Pal, U., Roy, P.P.: Multioriented and curved text lines extraction from indian documents. IEEE Trans. Syst. Man Cybern. Part B (Cybern.) **34**, 1676–1684 (2004)
11. Pal, U., Sinha, S., Chaudhuri, B.B.: Multi-oriented English text line identification. In: Bigun, J., Gustavsson, T. (eds.) SCIA 2003. LNCS, vol. 2749, pp. 1146–1153. Springer, Heidelberg (2003). https://doi.org/10.1007/3-540-45103-X_150
12. Setitra, I., Hadjadj, Z., Meziane, A.: A tracking approach for text line segmentation in handwritten documents. In: Proceedings of the 6th International Conference on Pattern Recognition Applications and Methods, ICPRAM 2017, Porto, Portugal, pp. 193–198, 24–26 Feb 2017
13. Stamatopoulos, N., Gatos, B., Louloudis, G., Pal, U., Alaei, A.: ICDAR 2013 handwriting segmentation contest. In: 2013 12th International Conference on Document Analysis and Recognition, pp. 1402–1406 (2013)
14. Khan, Z., Balch, T., Dellaert, F.: Multitarget tracking with split and merged measurements. In: 2005 IEEE Computer Society Conference on Computer Vision and Pattern Recognition (CVPR 2005), vol. 1, pp. 605–610 (2005)

Author Index

Printed in the United States
By Bookmasters